Health Impact Assessment

Health Impact Assessment for Sustainable Water Management

Edited by

Lorna Fewtrell and David Kay

Publishing

Published by IWA Publishing, Alliance House, 12 Caxton Street, London SW1H 0QS, UK
Telephone: +44 (0) 20 7654 5500; Fax: +44 (0) 20 7654 5555; Email: publications@iwap.co.uk
Web: www.iwapublishing.com

First published 2008

Typeset in India by Alden Prepress Services Private Limited
Cover design by www.designforpublishing.co.uk
Index provided by Christine Graham

British Library Cataloguing in Publication Data
A CIP catalogue record for this book is available from the British Library

Library of Congress Cataloging- in-Publication Data
A catalog record for this book is available from the Library of Congress

ISBN: 1843391333
ISBN13: 9781843391333

Contents

Preface

Health is increasingly, and rightly, being seen as a cross-cutting issue, which has important ramifications for many, if not most, development projects. Health impact assessment (HIA) conducted at an early stage in policy, programme or project implementation allows the explicit and timely consideration of possible health impacts and the development of strategies and recommendations for their mitigation.

This book on HIA for sustainable water management was born from a series of United Kingdom (UK) Engineering and Physical Sciences Research Council (EPSRC) projects and conversations with staff at the World Health Organization (WHO). The EPSRC projects focus on predicted health impacts and management option appraisal within a hypothetical UK-based population living on a new housing development, and an evaluation of health impacts resulting from flooding events. The UK research-centred case study chapters are contextualised by an examination of global approaches to water resource HIA development. This provides an interesting comparison of relative societal impacts in both socio-demographic settings.

This book was a multi-disciplinary enterprise, with authors from the fields of water science, public health, medicine, civil engineering, social science, epidemiology, microbiology, animal health, demographics, geography, social science and regional government.

The book aims to introduce the concepts, objectives and terminology of HIA and give an insight into the possible health impacts from the implementation of various water management developments, ranging from rainwater harvesting for non-potable use to irrigation projects and large hydroelectric projects and will be useful for water resource project designers, political analysts and international development professionals wishing to build an HIA component into new project proposals.

The editors are grateful to all the authors (and especially Jürg Utzinger and Robert Bos), who gave freely of their time and met the impossible deadlines with cheerfulness, enthusiasm and suggestions for other ways in which they could contribute.

There were many funding bodies which supported work resulting in the chapters presented below. Particular acknowledgement is due to the UK EPSRC for continued support for the development of HIA across a range of thematic programmes.

The opinions expressed in this publication are those of the authors and do not necessarily reflect the views or policies of their employers.

Lorna Fewtrell
David Kay

April 2008

Foreword

This year, 2008, it is 30 years ago that WHO and UNICEF were centre-stage on the public health scene with the Declaration of Alma Ata « Health for All by the Year 2000 », which laid the foundation for Primary Health Care (PHC). And today, under the inspiring leadership of Dr Margaret Chan, WHO is back as a stronger advocate of PHC than it has been at any time in the past 20 years.

A key pillar of the Health for All/PHC movement was and continues to be the need for intersectoral action for health: Health for All and All for Health. This makes a clear distinction between the responsibility for community health status ('everybody's business') and the responsibility for the delivery of health services, firmly anchored in the health sector. Intersectoral coordination, collaboration and action: loved by everyone, funded by no-one. Amazingly, nothing could be more appealing to our collective common sense, yet nothing has been so elusive ever since Alma Ata.

That may now fast become too pessimistic a view. There are increasingly strong indications that intersectoral thinking and action are taking hold in a sustainable way. Globally, it is significant that the Millennium Development Goals with their predominant explicit and implicit health targets were adopted at the level of heads of state, involving all government sectors equally and obliging them all to contribute and engage. At the national level, prevention of ill-health and health promotion are increasingly put on an intersectoral footing, whether it is through National Environmental Health Action Plans or through the implementation of the 2006 Bangkok Charter on Health Promotion in a Globalized World.

The growing importance of health impact assessment (HIA) fits well into this picture. Governments want to ensure that development leads to maximum benefits in quality of life with minimum ill effects on health and the environment. Project proponents not only want to reduce hazards and risks associated with their plans and actions, but they want to be seen to be doing so. And they want to enhance the sustainability of their development achievements. Public health authorities are eager to avoid the additional cost burden resulting from inadequately planned development and they wish to see overall socio-economic progress go hand in hand with improved health status.

Books about health impact assessment are not new. But in the past, they were largely written by health professionals and read by health professionals. This book establishes a new landmark: commissioned by the International Water Association, written in part by water professionals and intended for a broad readership of water professionals, managers, regulators and operators, it expands beyond conventional boundaries and brings information about HIA to professionals who will need to learn about its concepts, objectives and terminology and who will need to understand the value of this approach to their everyday decision-making. As such it is a valuable asset to the growing body of HIA literature.

Robert Bos
Scientist
World Health Organization
Public Health and Environment/Water, Sanitation and Health
Geneva, Switzerland

List of Contributors

Richard Ashley
Pennine Water Group, University of Sheffield, Sheffield, UK
Email: r.ashley@sheffield.ac.uk

Marci Z. Balge
NewFields, Denver, CO, USA
Email: mbalge@newfields.com

John Blanksby
Pennine Water Group, University of Sheffield, Sheffield, UK
Email: j.blanksby@sheffield.ac.uk

Robert Bos
Water, Sanitation and Health, Department of Public Health and Environment, World Health Organization, Geneva, Switzerland
Email: bosr@who.int

David Butler
Centre for Water Systems, University of Exeter, Exeter, UK
Email: d.butler@exeter.ac.uk

Soutsakhone Chanthapone
National Centre for Environmental Health and Water Supply,
Ministry of Health, Vientiane, Lao PDR
Email: soutch@laotel.com

Moses J. Chimbari
Harry Oppenheimer Okavango Research Centre, University of Botswana, Botswana
Email: mjchimbari@gmail.com

Jeroen H.J. Ensink
Environmental Health Group, London School of Hygiene and Tropical Medicine, London, UK
Email: jeroen.ensink@lshtm.ac.uk

Lorna Fewtrell
Centre for Research into Environment and Health, Aberystwyth University, Aberystwyth, UK
Email: lorna@creh.demon.co.uk

Peter Furu
Department of Veterinary Pathobiology, University of Copenhagen, Denmark
Email: pfu@life.ku.dk

Bruce Jefferson
Centre for Water Science, Cranfield University, Cranfield, UK
Email: b.jefferson@cranfield.ac.uk

Paul Jeffrey
Centre for Water Science, Cranfield University, Cranfield, UK
Email: p.j.jeffrey@cranfield.ac.uk

Surinder Kaul
Nam Theun 2 Power Company, Vientiane, Lao PDR and International SOS, Singapore
Email: surinder@namtheun2.com

David Kay
IGES, Aberystwyth University, Aberystwyth, UK
Email: dave@crehkay.demon.co.uk

Richard Kellagher
HR Wallingford Limited, Wallingford, UK
Email: r.kellagher@hrwallingford.co.uk

Flemming Konradsen
Department of International Health, Immunology and Microbiology,
University of Copenhagen, Denmark
Email: F.Konradsen@pubhealth.ku.dk

Gary R. Krieger
NewFields, Denver, CO, USA
Email: gkrieger@newfields.com

Ian Matthews
Centre for Health and Environment Research, Cardiff University,
Cardiff, UK
Email: matthewsip@cardiff.ac.uk

Adrian McDonald
School of Geography, University of Leeds, Leeds, UK
Email: a.t.mcdonald@leeds.ac.uk

Fayyaz Ali Memon
Centre for Water Systems, University of Exeter, Exeter, UK
Email: f.a.memon@exeter.ac.uk

Peter Odermatt
Department of Public Health and Epidemiology, Swiss Tropical Institute,
Basel, Switzerland
Email: peter.odermatt@unibas.ch

Pany Sananikhom
Nam Theun 2 Power Company, Vientiane, Lao PDR
Email: pany@namtheun2.com

Adrian Saul
Pennine Water Group, University of Sheffield, Sheffield, UK
Email: a.j.saul@sheffield.ac.uk

Burton H. Singer
Office of Population Research, Princeton University, Princeton, NJ, USA
Email: singer@princeton.edu

Keren Smith
Centre for Research into Environment and Health, Aberystwyth University, Aberystwyth, UK
Email: keren@netcase.co.uk

Marcel Tanner
Department of Public Health and Epidemiology, Swiss Tropical Institute, Basel, Switzerland
Email: marcel.tanner@unibas.ch

Jürg Utzinger
Department of Public Health and Epidemiology, Swiss Tropical Institute, Basel, Switzerland
Email: juerg.utzinger@unibas.ch

Wim van der Hoek
Bierstalpad 37, 1121 JK Landsmeer, Netherlands
Email: w.vanderhoek@compaqnet.nl

John Watkins
CREH Analytical, Leeds, UK
Email: john@crehleeds.demon.co.uk

Abbreviations

ACTs	Artemisinin-based combination therapies
AIDS	Acquired immune deficiency syndrome
AUDACIOUS	Adaptable urban drainage – addressing change in intensity, occurrence and uncertainty of stormwater
CA	Concession agreement
CDC	Centers for Disease Control and Prevention
cfu	Colony forming unit
cpu	Cytopathic unit
CI	Confidence interval
CSO	Combined sewer overflow
CSIP	Care Services Improvement Partnership
DALY	Disability adjusted life year
DBP	Disinfection by-products
DEFRA	Department for Environment, Food and Rural Affairs
DOTS	Directly observed therapy, short course
DTI	Department of Trade and Industry
EA	Environment Agency
EHA	Environmental health area
EIA	Environmental impact assessment
EPFI	Equator principles financial institutions
EPSRC	Engineering and Physical Sciences Research Council
FAO	Food and Agriculture Organization
FHRC	Flood Hazard Research Centre
FRMRC	Flood risk management research consortium
GAD	Government's Actuary Department

GHQ-12	General health questionnaire (12-item)
GoL	Government of Lao
GP	General Practitioner
HASS	Home accident surveillance system
HCGI	Highly credible gastrointestinal symptoms
hh	Households
HIA	Health impact assessment
HIV	Human immunodeficiency virus
HPA	Health Protection Agency
HPP	Healthy public policy
HSE	Health and Safety Executive
IAIA	International Association for Impact Assessment
IFC	International Finance Corporation
IPCC	Intergovernmental Panel on Climate Change
IRR	Incidence rate ratio
ITN	Insecticide-treated nets
LASS	Leisure accident surveillance system
MBR	Membrane bioreactor
MIP	Mupfure irrigation project
MoH	Ministry of Health
MoU	Memorandum of understanding
MPN	Most probable number
NEPA	National Environmental Protection Act
NERC	Natural Environment Research Council
NTPC	Nam Theun 2 Power Company
OR	Odds ratio
p	Probability
pcc	Per capita consumption
PDR	People's Democratic Republic
PEEM	Panel of Experts on Environmental Management for Vector Control
pfu	Plaque forming unit
PHC	Primary health care
PIAOC	Potential impact area of concern
PM	Particulate matter
ProgMan	Programme management
PTSD	Post Traumatic Stress Disorder
PTSS	Post Traumatic Stress Scale
QMRA	Quantitative Microbial Risk Assessment
Qual	Qualitative

RCD	Residual current device
RoSPA	Royal Society for the Prevention of Accidents
RR	Relative risk
RWH	Rainwater harvesting system
SEA	Strategic environmental assessment
SF-12	Short form (12 item)
SFPHM	Scottish Forum for Public Health Medicine
SIA	Social impact assessment
STD	Sexually transmitted disease
SUDS	Sustainable drainage
TB	Tuberculosis
UKCIP	UK Climate Impacts Programme
UKWIR	UK Water Industry Research
UNEP	United Nations Environment Programme
USDHHS	United States Department of Health and Human Services
UV	Ultraviolet
VFCW	Vertical flow constructed wetland
VGDHS	Victorian Government Department of Human Services
VKT	Vehicle kilometres travelled
WaND	Water cycle management for new developments
WASA	Water and Sanitation Agency
WCD	World Commission on Dams
WHO	World Health Organization
WNV	West Nile virus
WSP	Waste stabilization pond
Wwd	Water-washed diseases
WWTW	Wastewater treatment works
YLD	Years lived with disability
YLL	Years of life lost

1

Health impact assessment for sustainable water management: the lay of the land

Lorna Fewtrell, David Kay, Ian Matthews, Jürg Utzinger, Burton H Singer and Robert Bos

1.1 BACKGROUND

Health impact assessment (HIA) aims to predict public health issues 'upstream' and hence to influence the decision-making process in development planning (Kemm and Parry 2004). This facilitates the formulation of healthy public policy (HPP), one of the conceptual roots of HIA, and the incorporation of measures for health protection and promotion in proposed policy, projects and programmes. Environmental impact assessment (EIA), with a 35+ year history, is seen as the second conceptual root of HIA (Kemm and Parry 2004). Over the past 20 years,

 Health Impact Assessment for Sustainable Water Management. Edited by Lorna Fewtrell and David Kay. ISBN: 9781843391333. Published by IWA Publishing, London, UK.

innovation, validation and application of HIA procedures, methods and tools have accelerated, and a number of countries have adopted national HIA policies and strategies. While these countries are mainly found in the industrialized world, recently some of the least developed countries have also adopted HIA policies (e.g. Lao PDR adopted its national HIA policy in March 2006).

In the past, development was generally undertaken without even an informal assessment of possible health impacts. This prior lack of assessing possible health impacts can be explained by the fact that health was considered the exclusive responsibility of the health sector, a view that still prevails in some countries or sectors. As a result, a range of adverse health effects added a preventable burden of ill-health to the lives of vulnerable population groups. Most development also missed opportunities for health promotion that typically present themselves at project planning and design stages. Unintended negative health effects have included such diverse impacts as the spread and intensification of vector-borne disease transmission, particularly schistosomiasis and malaria associated with hydropower projects and irrigation systems (Southgate 1997; Jobin 1999; Scudder 2005; Steinmann *et al.* 2006), intensification of dengue transmission due to the introduction of storage jars for drinking water in Viet Nam (Vu *et al.* 1998), arsenic poisoning as a result of using contaminated boreholes for supplying drinking water in Bangladesh (Smith *et al.* 2000) and the psycho-social disorders resulting from forced resettlement out of reservoir areas or irrigation schemes (Montgomery *et al.* 1973; Scudder 2005). An example of missed health opportunities in developing countries is that of numerous irrigation schemes developed without including a component to improve access to safe drinking-water and sanitation (Konradsen *et al.* 1997).

A major feature of large development projects in developing countries (e.g. the construction and operation of hydroelectric power schemes, mining projects and oil and gas pipelines) is involuntary resettlement of entire communities. The number of people involved in involuntary resettlement is estimated to be in the region of ten million persons per year (Goodland 2000). The negative health consequences of such resettlement have been documented in studies dating back 50 years (Scudder 2005) and recommendations that rigorous appraisal of potential health impacts be carried out in the context of resettlement linked to large dams was proposed in 1973 (Montgomery *et al.* 1973) with repeated admonition to build this into the planning stages of large projects. However, the term HIA was only introduced in the scholarly literature many years later and no systematic follow-up of HIAs in the important context of large dams has been forthcoming. Figure 1.1 underscores the magnitude of the issue, showing the number of large dams constructed in different regions of the world over the past 100 years. Of note, the People's Republic of China is not included in these statistics, although it is estimated that over 50% of the large dams have been constructed in this country alone.

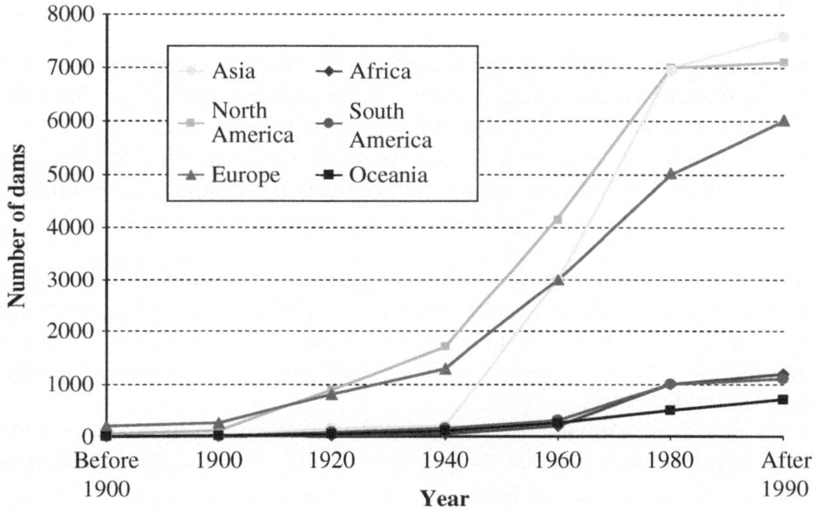

Figure 1.1. Cumulative number of large dams constructed in different regions of the world over the past 100+ years (excluding the People's Republic of China)

When HIAs have been implemented in connection with large projects in the tropics and subtropics, primary attention is often focused on infectious disease, with little or no consideration given to what is probably the most profound long-term health consequence of resettlement; namely, mental illness (Scudder 2005). Furthermore, HIAs alone cannot deal with long-term health consequences of development projects. If large infrastructure development projects are not accompanied by an ongoing data collection programme and public health management plan capable of being adaptively tuned to local conditions over time, then even the best of HIAs will have limited impact. Discussion of this topic in the context of wealthier countries focuses on government responsibility and legally binding agreements for promotion of health of the public. In most tropical and subtropical regions of the world, there is no legal system with this kind of focus, Brazil being a notable exception, and project development usually involves a three-way tension between international development banks, corporations and host governments. If ongoing health monitoring and implementation of curative and preventive services is not actually written into a concession agreement (CA) between the corporations and the government, there is little hope that HIAs will be anything other than consciousness raisers at the start of a project. Furthermore, addition of health mitigation into the formal CA assumes that:

- some type of HIA was performed prior to the CA negotiations;
- sufficient capacity exists within the host government for long-term implementation of the required mitigation measures; and
- a monitoring and evaluation system of sufficient capacity has been developed in order to identify and evaluate both impacts and mitigation measures.

Basically, HIAs should not be treated in isolation from the long-term follow-up that is needed to mitigate health consequences of development projects. The pressing need for such follow-up is already documented in the analysis of over 50 resettlement cases connected to large dams, many running over multiple decades (Scudder 2005).

With increasing demands for 'sustainable development' (Goodland 1995), and the acknowledgement that the greatest scope for improving public health lies outside the traditional medical profession (which focuses on management of illness) (Lock 2000), HIA has become an important instrument to manage and mitigate possible health implications in advance of any development. Consequently, HPP options can be selected and an intersectoral public health management plan can be integrated into the design, implementation and operation of development projects and programmes. The principles of sustainable development have been summarised as:

- environment;
- equity (i.e. within and between societies); and
- futurity (i.e. inter-generational equity).

Achievement of all three elements in the absence of a strong public health ethos is impossible and this explains the emerging focus on the health dimension in sustainable development policy goals (WHO ECHP 1999; Kemm and Parry 2004; Petticrew *et al.* 2004).

HIA can provide an estimation of potential health impacts (positive or negative; intended or not; direct or indirect; single, multiple or cumulative). While HIA is predominantly qualitative, for certain issues it is possible to be quantitative with varying levels of uncertainty (Veerman *et al.* 2005). It can be used to define health safeguards, mitigation measures and health promotional activities for the design, construction and operational project phases (Bos 2006; Chapter 9). Thus, HIA supports minimization of predicted health risks and makes it possible to take optimal advantage of health opportunities.

The enthusiasm with which HIA has been adopted varies. The United Kingdom is a leading nation, certainly in terms of published reports in the peer-reviewed international literature. Figure 1.2 is reproduced from a systematic review covering the period from 1976 to mid-2007 which revealed that almost a quarter of HIA-related publications in the peer-reviewed literature focused on the United Kingdom (Erlanger *et al.* 2008). This imbalance explains the difficulty in sourcing developing country HIA examples for a book of this nature. This issue was noted when critically reviewing a recent textbook on HIA (Kemm *et al.* 2004); indeed, only one of the 36 chapters has an explicit focus on developing countries and all of the 55 contributing authors are affiliated to high-income nations (Utzinger 2004).

Figure 1.2. HIA-related publications in the peer-reviewed literature from 1976 to mid-2007, stratified by country (reproduced, with permission, from Erlanger *et al.* 2008)

Many countries, however, do have policies and legislation regulating EIA, and such assessments have traditionally included a health component. The scope of health coverage in EIAs tends to reflect its roots in industrialized countries in the 1960s and 1970s, with a focus on pollution, chemical hazards, accidents and injuries (Steinemann 2000). In the United States of America, despite more than a 30 year history of formal EIA legislation, HIAs are only now beginning to be incorporated into the National Environmental Protection Act (NEPA) and relatively few HIAs have actually been performed for any industrial development projects. Furthermore, EIA tends to focus on the physical environment with little or no attention to the cultural, economic and societal aspects of development projects which include equity, health, livelihood and wellbeing. There may also be conflicts of interest between the environment and health sectors (an extreme example would see wetland conservation juxtaposed to wetland drainage for malaria or dengue vector control), and the boundaries set for the scope of an EIA may differ significantly from those required for a parallel HIA.

Kemm and Parry (2004) identify two key features of an HIA, namely:

- it is concerned with the prediction of different options and how these contemplated decisions would affect the health of a population; and
- it is intended to influence decision-making and to assist decision makers.

Therefore, by definition, an HIA needs to be undertaken prior to the implementation of a policy, project or programme. Thus, HIA is predictive. A formidable challenge of HIA with regard to influencing the decision-making process is the limited time available for such development decisions and the notorious lack of high-quality baseline health and socioeconomic data, particularly in developing country settings. One suggestion is for a number of 'off-the-shelf' reviews (Joffe and Mindell 2002; Mindell *et al.* 2004) to speed up subsequent HIA. The United Kingdom research reported in Chapters 3 to 5 is, in part, a response to this. These desk-based HIAs have been conducted on a hypothetical case-study population.

This Chapter does not set out to reinvent the wheel; there is already good guidance available on how to manage and how to conduct an HIA in a relatively wealthy country setting, including information on suitable stakeholder groups and public engagement (see for example enHealth 2001; Abrahams *et al.* 2004). Rather, it sets the scene so that the approaches taken in the research reported in subsequent chapters can be contextualised. For those who wish to perform their own HIAs on water management options in the United Kingdom (such as utilising rainwater harvesting and greywater reuse), specific guidance is available (Fewtrell and Kay 2008) in addition to the case study examples outlined in subsequent chapters. In a developing country setting, with the exception of HIA guidelines that have been developed for the oil and gas industry (IPIECA 2005) as a derivative of the HIA carried out for the Chad-Cameroon petroleum development and pipeline project (Jobin 2003; Utzinger *et al.* 2005), there is little of a general nature to build on that derives from concrete project experience. Lessons from the Nam Theun 2 hydroelectric project in Lao PDR (Chapter 9) represent one place from which to make further progress. Some of these lessons are under consideration in guidelines for HIAs being developed through the International Finance Corporation (IFC)[1].

1.2 ORIGINS OF HIA

Despite its widespread uptake in many industrialized countries, and despite strong pleas made in the 1970s for the need to assess health impacts in the context of large dams and involuntary human resettlement in developing countries (Montgomery *et al.* 1973), HIA only really emerged in the 1980s. The trigger was a specific concern over the vector-borne disease implications of water resources development and management, specifically irrigation schemes and dams implemented in developing countries. To address this issue, the World Health Organization (WHO), Food and Agriculture Organization of the United Nations (FAO) and United Nations Environment Programme (UNEP) established a joint Panel of Experts on Environmental Management for Vector Control (PEEM). One of the first panel meetings addressed how vector-borne disease implications of water resources development could be predicted (WHO 1983). Recommendations made at this meeting were followed up with the development of forecasting guidelines, published in their final version in 1991 (Birley 1991).

Despite this chain of meetings, there is a considerable distance between guidelines developed in the context of international organizations and the actual conduct of HIAs on specific projects with a focus on developing countries. Here the HIA experience for the Chad-Cameroon petroleum development and pipeline project (Jobin 2003; Utzinger *et al.* 2005), the Ahafo gold mining site in Ghana and the Nam Theun 2 hydroelectric project in Lao PDR (Chapter 9) provide a basis for going well beyond broad guidelines and moving to concrete protocols derived from, and improving upon, what has actually been done in the complex triangle of the corporate world, governments and international development banks, that is at the heart of large projects in the tropical regions of the world.

It is also noteworthy that during the 1980s, there was a rapidly expanding interest in EIA, nurtured by a series of articles in "*The Ecologist*" on the impacts of dams and other major infrastructure development projects in low-income

[1] The IFC HIA 'toolkit' is due to be released in mid-2008. IFC did update its 'Guidance Notes for Community Health' impacts in 2007 and this revised and expanded material is available at the IFC website (http://www.ifc.org).

countries. Indeed, there was a growing recognition that, despite the requirement for a consideration of people's health and wellbeing within EIA (NHMRC 1994; Birley 1995; BMA 1998), health only received token attention (Arquiaga *et al.* 1994; NHMRC 1994; Birley 1995). At the same time, there was an interest in HPP. A landmark event was the 1986 Ottawa Charter on Health Promotion, which spelled out the critical importance of HPP as part of intersectoral approaches to health (Ottawa Charter 1986). The role that HIA could play was realized and enabled through the publication of guidance (PHRB 1994; Putters 1994; DoH 1995). The Bangkok Charter (2006) re-iterated this, with explicit reference to the key-role of equity-oriented HIA.

1.3 DEVELOPING RATIONALE

HIA is still at a relatively early stage of development in comparison with other types of impact assessment, most notably EIA. There is a need, therefore, to strengthen the role of health in various impact assessment processes, such as EIA and strategic environmental assessment (SEA) (Bina 2007), as well as to encourage the development of HIA in its own right. Linked to this, there is also a need to develop criteria to decide on the level of integration between HIA and other types of impact assessment. The bottom line is that further progress on HIA methodologies and tools, underlying processes and integration with other impact assessments will only come about by doing more HIAs (Parry and Kemm 2004; Utzinger 2004). The pressing need of HIA in developing country settings, particularly in the frame of large-scale water resources development and management projects and exploration of other natural resources has been stressed (Erlanger *et al.* 2008).

Investing in HIA is justified by four compelling arguments: economics, equity, best public health practice and good governance, all of which are required for policy implementation based on principles of sustainable development (Bos *et al.* 2003).

1.3.1 Economics

The economic argument is perhaps the strongest driver for ministries of finance, developers, local agencies, donor agencies and staff in sectors other than health to engage in HIA. Past experience shows that a lack of consideration of possible health issues in development may result in the transfer of hidden costs to the health sector, which will have to deal with new or increased disease burdens in affected communities. As a consequence, the savings perceived by the project proponent resulting from not doing an HIA are false economies when the issue is considered government-wide and the value of the additional burden on the health sector is included in the analysis. This is clear-cut in some examples, such as the malaria outbreaks in the Gezira irrigation scheme in Sudan, where the army had to be called in to save the cotton harvest, because all agricultural labour was down with irrigation-associated malaria (Bell 1999). Or, in the example of the Accelerated Mahaweli Development Project in Sri Lanka, where 165,000 ha of land were developed for irrigated rice farming accompanied by the settlement of approximately 1 million individuals into the malaria-endemic lowland dry zone in the 1980s. There was a 2- to 5-fold increase in malaria incidence. In one part of the irrigation scheme, the unfortunate combination of intensified irrigated rice production and the promotion of pig rearing, as a second source of income, further led to severe outbreaks of Japanese encephalitis, which forced the government to embark on costly vaccination campaigns (Amerasinghe 2003). Another prominent example is the severe outbreak of intestinal schistosomiasis in the Richard Toll area of the Senegal River basin in connection with the construction and operation of the Diama dam and intensified irrigation (Southgate 1997). Although some kind of HIA had been performed, the risk of schistosomiasis had been underestimated as insufficient attention had been placed on the ecology and possible spread of the intermediate host snails of *Schistosoma mansoni* in this area.

More often, however, the impacts, and the burden development projects place on the health sector, are more insidious, and the association between development and ill-health may be less obvious. There may be various pathways from hazard to clinical illness, or the timescale between cause and effect may be so large that confounding factors become significant. Yet, the bottom line is that the burden on the health services has a chance of being managed effectively by adequate HIA upfront in the project development cycle, thus reducing the potential that developers simply shift a cost for ill-health to the health service, in effect, avoiding a major externality of the development process. Moving from potential to effective longer term follow-up will require agreements between corporations and governments that have real force behind them. One route into this is to have health follow-up written into the CA that is articulated and signed early in the project planning process. There is only minimal, and decidedly inadequate, consideration of health issues in CAs to-date. However, since action by both the corporations and governments in all aspects of project development is driven by the CA this is where pressure must be brought to bear if HIAs are to be conducted in a timely fashion and influence project planning.

As an example of this phenomenon, it is useful to first observe that a series of health checks, with accompanying reports are being carried out for the Nam Theun 2 hydroelectric project under the auspices of the Nam Theun 2 Power

Company (NTPC) because it is included in the CA. While an HIA was carried out for this project in early 2004 (Chapter 9), it was not as a consequence of the CA. Indeed, the order of events was reversed. The CA was signed and later there was an HIA with a follow-up public health management plan. While it has been projected that communities in the Xe Bang Fai areas will be impacted following closure of the dam on the Nam Theun river and discharge of the water into the Xe Bang Fai hydrological system, less attention is given to these communities than the resettled people on the Nakai plateau. One of the underlying reasons is that requirements spelled out in the CA regarding the Xe Bang Fai areas are less explicit. A further point is that there were tight time constraints for the conduct and reporting of the HIA of the Nam Theun 2 hydroelectric project (roughly six weeks), which makes it difficult for quantitative projections that should form part of a comprehensive HIA.

In general, there is an opportunity cost associated with not taking health into account upstream in the development process: frequently, the most cost-effective public health measures can only be deployed at the design and construction phases of large development projects. These health opportunities often comprise modifications in design or additional infrastructural and engineering features, allowing operational rule curves[2] that support health protection. They may be unavailable as development gets under way, the concrete starts being poured and the options for modifications to project design and operational management are reduced. As ill-health effects become apparent, the health sector will have to deploy recurrent, often less cost-effective and essentially remedial interventions. Strikingly, for example, many irrigation schemes implemented in the 1980s and 1990s did not have a drinking-water supply and sanitation component for the local communities, in spite of the fact that the efficiency of including such a component at the design stage would be significantly greater than implementing it separately after the irrigation project had become operational (Tiffen 1991). Lack of intersectoral coordination, or sometimes poor economic judgement (e.g. concern that a drinking-water component would negatively affect the project's internal rate of return), was often at the root of this problem. Parallel professional 'silo' working has been evident in the developed nations where health concerns have often been raised by activist groups opposing development rather than being addressed as a central component of the planning system (Bond *et al.* 2002). It is precisely this generic problem that HIA seeks to overcome and, here, the legal commitment to sustainable development

[2] Guidance on operation draw down to provide hydroelectric power and water supply, while retaining downstream river flow.

principles by the regulatory communities in developed and developing nations provides the most significant supporting rationale.

Finally, still from an economic perspective, a healthy human resource base is essential for sustainable development. Fortunately, there have been few projects as dramatic as the Bura Irrigation Project in the Upper Tana River valley in Kenya (Sir M. MacDonald and Partners *et al.* 1977), which initially had to be abandoned because of high malaria mortality among non-immune children and adults who had been resettled from highland areas. Yet, in many instances the development investments have not yielded results to their full potential because the human resource base was undermined by ill-health. In the words of former WHO Director-General Dr Halfdan Mahler: "*I know that health is not everything, but without health, everything else is nothing. When human health in development is overlooked, it lacks the community's physical energy, its spiritual energy*" (Birley 1991).

1.3.2 Equity

The equity argument focuses on vulnerability. Development, by definition, aims to produce benefits for communities, increase their quality of life and enhance the security of their livelihood. In the case of water resources development and management in low-income countries, for example, elimination of floods and meeting important economic and political needs are portrayed as major benefits. Yet, the distribution of these benefits and, indeed, of unintended health risks, may show significant variation within the overall community. Large-scale hydroelectric projects and irrigation schemes are typically superimposed upon a local community which has virtually no say during the early feasibility studies, planning phases and actual implementation of the project (Montgomery *et al.* 1973). Thus, the potential for uneven differences is a key issue in HIA, and hence it is important to focus on the vulnerable population groups, those whose health status may be most tangibly affected by proposed development.

We suggest that the traditional risk assessment and mitigation issues of an HIA could be further developed by taking into consideration a vulnerability and resilience framework (deFur *et al.* 2007; Obrist *et al.* 2007). In doing so, it will be important to deepen our understanding of the vulnerability context of communities that will be affected by large water resources development projects. People's vulnerability context is governed by a host of factors over which they have little influence, e.g. economy, politics, technology, climate variability and extreme events such as armed conflict and war, droughts and floods. Within a given vulnerability context, access to livelihood assets by individuals, households, communities and the society at large then shapes the recognition of an illness and subsequent health-seeking behaviour. Since livelihood assets comprise human, social, natural, physical and financial capital, these issues need to be understood in order design sound mitigation strategies, according to potential resilience strategies of local communities already in place.

As the constitutional and treaty articles below clearly indicate, equity is a key attribute, as is plural democracy, in the sense that the majority in society looks after the vital interests of minority groups. In some countries, the right of all citizens to demand an advance scrutiny of development has been laid down in the Constitution, as is the case in Thailand. Article 57 of the Thai Constitution reads: "*A person shall have the right to receive information, explanation and justification from a government agency, State agency, State enterprise or local government organization* ***before permission is given for implementation of any project or activity*** *which may affect* ***the quality of environment, health, and sanitary conditions, quality of life*** *or other material interest to him or her or a local community and shall have the right to express his or her opinion on such matters to agencies concerned for their consideration*". And Article 129 of the 1999 EU Treaty of Amsterdam reads: "*A high level of human health protection shall be ensured in the definition and implementation of all Community policies and activities*".

Fundamental, at the start of any HIA, is the identification of vulnerable groups. Vulnerability is linked to environmental and social determinants of health (see section 1.5.2), and it is the development-induced changes in these determinants that may further enhance vulnerability. Seemingly irrelevant changes, such as the changes in water flow regimes in downstream areas, may have repercussions for health. In developing countries, increased unreliability of water availability may hamper subsistence agriculture and lead to malnutrition, while increased water availability may encourage intensified agriculture and associated use of pesticides, increasing exposure risks to these hazardous chemicals. HIA holds promise to readily identify vulnerable groups and to protect them from negative impacts taking into consideration their access to livelihood assets.

While it is acknowledged that development brings many benefits (including health benefits) to large groups of people, there can be no trade-off's between this and the adverse impacts on vulnerable communities. HIA aims to ensure that exactly these communities receive optimal protection in the context of policies, projects and programmes.

While these principles are often illustrated with developing country examples, equity and vulnerability concepts are just as applicable to industrialized nations where interventions to enhance sustainable development can have disproportionate psychological and, potentially, physical health impacts on the poorest members of society. This was examined in relation to potable water demand management through the introduction of domestic water metering and resultant supply disconnection policies by Fewtrell *et al.* (1994) in the United Kingdom.

1.3.3 Public health

'Impact' as a term has developed negative connotations. The public health argument for HIA emphasizes, however, both positive and negative impacts. Development affects hazards as well as promoters of health, and results in a complex array of evolving determinants of health, with outcomes that can be negative or positive. Clearly, the communities resettled from the Nakai plateau in the Nam Theun 2 hydroelectric project in Lao PDR (Chapter 9) will have considerable health benefits in terms of reduced exposure to agents of communicable diseases (such as the micro-organisms in contaminated drinking-water causing diarrhoeal diseases), but the promotion of more intensive irrigated subsistence agriculture and close proximity to a large water reservoir may also increase the risk of disease transmission by water-associated vectors. It is, thus, critically important that the terms of reference for an HIA give instructions to identify the health opportunities and associated health promotional measures to achieve them.

The links between HIA and health promotion are multiple (Bos 2006) and any public health management plan should have a substantive component dealing with health education and other health promotional measures.

1.3.4 Good governance

Finally, HIA is compatible with, and reinforces, best practice in governance. By nature, it requires different sectors to collaborate in a common framework. Like EIA, HIA champions public participation, and in some countries it has strong subsidiarity characteristics, i.e. decisions are made at the lowest possible level, with higher administrative levels only becoming involved when issues cannot be resolved at the local level. A number of other characteristics, which are described in detail in the Report of the World Commission on Dams (WCD 2000) are also closely associated with HIA, such as: consideration of voluntary versus involuntary risks and related criteria; mechanisms and procedures for compensation of harm done; or compliance performance contracts with all stakeholders, with a strong (health-) protective component towards local communities and the possibility to apply sanctions for non-compliance.

In brief, HIA is comprehensive in its way of considering impacts, adds economic and social value to development, fosters equitable and sustainable development and, as a member of the impact assessment family, helps to determine sound development options.

1.4 SUSTAINABLE WATER MANAGEMENT

In many countries, the pressure on water resources is intensifying, with water scarcity already affecting every continent and four out of every ten people in the world (Rosegrant and Cline 2003; Moe and Rheingans 2006). Population growth and urbanization will exacerbate this problem (WHO 2007). Thus, there is a need to develop new resources, decrease demand, enhance efficient water use and implement recycling or reuse measures.

The public and private sectors responsible for water resources development and management operate under a great deal of pressure.

- Sustainable utilisation of the water resource whether surface or groundwater with no 'mining' of geogenic waters and appropriate treatment of 'return' flows to prevent ecosystem damage in receiving waters and ensure resource availability for downstream communities.
- Equity in access to sufficient clean water for all members in society regardless of their individual economic and/or political power.
- A long-term, multi-generational, perspective on resource use to ensure continued availability despite projected socio-demographic and agro-industrial developments.
- Communication between the three elements above to ensure integrated and sustainable resource development.

1.5 DEFINITIONS OF HEALTH AND HIA

1.5.1 Health

The definition of health adopted greatly affects the subsequent HIA. There are two broad paradigms:

- the reductionist 'biomedical' approach, where health is considered within a series of disease categories (e.g. communicable disease, non-communicable disease, nutrition, injury, mental disorder, etc.) and the health sector structure and operations deal with these through the delivery of health services; and

- the 'social or public health model' where health is considered to be "*a state of complete physical, mental and social well-being and not merely the absence of disease or infirmity*" (WHO 1946).

A range of complex factors (determinants) are acknowledged to play a role in community health and it is these determinants that are affected by development decisions. HIA aims to predict the changes in the determinants.

1.5.2 Determinants of health

The broad definition of health adopted by the WHO can be illustrated by considering the determinants of health as outlined in Table 1.1. From this Table it can, however, be seen that virtually every area of human activity influences health.

Table 1.1. Example determinants of health (adapted from enHealth 2001)

Determinant	Examples
Fixed	Genes Gender Ageing
Social and economic	Poverty Employment Social exclusion Community structure and infrastructure
Lifestyle and behaviour	Diet Physical activity Smoking Alcohol Sexual behaviour Drugs Coping skills
Access to services	Education Health services Social services Transport Leisure
Environment	Air quality Noise Housing Water quality Social environment Risk of injury Sun exposure Disease vectors

In the systematic approach used in an HIA, the existing health hazards and health promoters are identified, as well as new hazards/promoters that may be introduced into an area as a result of a project, programme or policy.

Changes in the determinants of health caused by development have an impact on the probability (i.e. risk, or in the case of health promoters: opportunity) that the hazard or promoter leads to a change in health status. Individual and community vulnerability, environmental receptivity (both physical and social environment) and the capacity of the health services are the main categories of determinants under which items from Table 1.1 can be grouped (as illustrated in Chapter 8). Thus, a matrix can be composed that sets of hazards against determinants and provides qualitative insights as to whether or not, for each hazard, risk levels will be increased, reduced or remain the same (an example is put forward in Chapter 9).

1.5.3 HIA

Impact assessment has been defined by the International Association for Impact Assessment (IAIA 2007) as follows: "*Impact assessment, simply defined, is the process of identifying the future consequences of a current or proposed action*". There are numerous definitions of HIA (some more prescriptive than others), as illustrated in Table 1.2, and probably as many ways of performing the actual assessment, and it is widely accepted that there is "*no single 'blueprint' for HIA that will be appropriate in all circumstances*" (Douglas *et al.* 2001).

In addition to prompting the maximum health of the population, four values are particularly important for HIA (WHO ECHP 1999; Kemm and Parry 2004):

- **Democracy**, emphasizing the right of people to participate in a transparent process for the formulation, implementation and evaluation of policies that affect their life, both directly and through the elected political decision makers.
- **Equity**, emphasizing that HIA is not only interested in the aggregate impact of the assessed policy on the health of a population but also on the distribution of the impact within the population, in terms of gender, age, ethnic background and socio-economic status.
- **Sustainable development**, emphasizing that both short- and long-term as well as direct, indirect and cumulative impacts are taken into consideration.
- **Ethical use of evidence**, emphasizing that the use of quantitative and qualitative evidence has to be rigorous, and based on different scientific disciplines and methodologies to get as comprehensive assessment as possible of the expected impacts.

Table 1.2. Definitions of HIA (adapted from Krieger *et al.* 2003)

Definition	Source
"A combination of procedures, methods and tools by which a policy, programme or project may be judged as to its potential effects on the health of the population and the distribution of those effects within the population"	WHO ECHP 1999
"A means of evidence based policy making for improvement in health. It is a combination of methods whose aim is to assess the health consequences to a population of a policy, project, or programme that does not necessarily have health as its primary objective."	Lock 2000
"A multi-disciplinary process within which a range of evidence about the health effects of a proposal is considered in a structured framework, … based on a broad model of health which proposes that economic, political, social, psychological and environmental factors determine population health"	NYPHO 2001
"Can best be described as a decision-making tool, one that is designed to take account of the wide range of potential effects that a given proposal may have on the health of its target population. Thus, it is a process that: considers the scientific evidence about the relationships between a proposed policy, programme or project and the health of a population; • takes account of the opinions, experience and expectations of those who may be affected by a proposed policy decision;	National Health Service Executive 2001

Table 1.2. *Continued*

Definition	Source
• highlights and analyses the potential health impact of the proposed policy decision; • enables decision makers to make fully informed decisions and to maximise positive and minimise negative health impacts; and • enables consideration of health inequalities."	
"Is a developing approach that can help to identify and consider the potential – or actual – health impacts of a proposal on a population. Its primary output is a set of evidence-based recommendations geared to informing the decision making process."	Taylor and Quigley 2002
"Provides a structured framework to map the full range of health consequences of any proposal, whether these are negative or positive. It helps clarify the expected health implications of a given action, and of any alternatives being considered, for the population groups affected by the proposal. It allows health to be considered early in the process of policy development and so helps ensure that health impacts are not overlooked."	WHO 2002
"A combination of procedures, methods and tools that systematically judges the potential and sometimes unintended effects of a policy, plan, programme or project on the health of a population and the distribution of those effects within the population. HIA identifies appropriate actions to manage those effects."	IAIA/WHO 2006

1.6 THE HIA PROCESS

The HIA process consists of procedures (a sequence of steps that need to be systematically applied and that are overseen by a regulator) and methods (for each step, specific methods have been developed that can be applied in relation to the specific conditions for a given HIA). In brief: procedure is about management and method is about technique, which may include screening checklists and frameworks. The basic steps of the procedure (Ewan *et al.* 1994; HDA 2002; Kemm and Parry 2004) include:

- **Screening**, a desk-based preliminary exercise applied to a proposed project, programme or policy, to decide whether investing in a full-scale HIA is warranted.
- **Scoping**, a process setting the temporal and spatial boundaries for the HIA, identifying the range and types of negative and positive health effects considering specific phases of the project and identifying vulnerable population groups. The output of the scoping process can be utilized as a basis for a formal set of terms of reference for the full HIA.
- **Assessment**, the actual hazard identification, risk assessment and impact synthesis process, followed by the formulation of conclusions and recommendations for safeguards, mitigating measures and health promotion (normally carried out by consultants). A key issue for HIAs carried out in developing country settings is the lack of quality baseline health data.
- **Appraisal**, the quality control component of HIA, focusing on the objectivity, comprehensiveness and evidence base of the HIA and on the technical soundness, social acceptability and economic feasibility of proposed measures – normally carried out by the government regulator in the Ministry of Health.
- **Formulating and prioritising recommendations**, consolidation of recommendations into a public health management plan, with well founded arguments to obtain the necessary resources.
- **Negotiation**, with the Ministry of Finance over allocation of funds for the different components of the public health management plan and with the proponent and other relevant line ministries over the institutional arrangements determining roles and responsibilities for the plan's implementation;
- **Implementation of the public health management plan**, the key activity is monitoring of compliance with the agreed plan, of the health status of affected communities and of the effectiveness of proposed measures.

It is rare for HIA to report on the screening process as it is taken as read that if a full HIA has been conducted the initial proposal passed successfully through the screening process. A screening procedure, outlining useful questions is shown in Figure 1.3 (Cole *et al.* 2005).

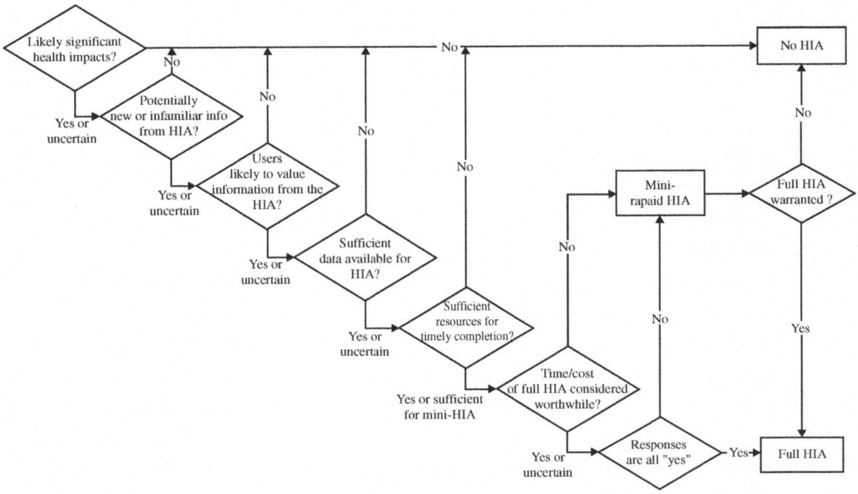

Figure 1.3. Screening process to guide decision on whether to conduct a HIA (reproduced, with permission, from Cole *et al.* 2003)

The majority of the subsequent chapters focus on the assessment step of the HIA process.

1.6.1 Levels of HIA

HIAs may vary in depth and complexity as illustrated by Table 1.3. Although the depth and complexity of HIAs have been neatly categorised into levels in Table 1.3, in reality each individual HIA lies somewhere on a continuum, with aspects and methodologies combined to provide the most appropriate assessment given time and financial constraints.

The emphasis in the United Kingdom-based research (Chapters 2 to 7) was on a desk-based approach, while those illustrating developing country HIAs (Chapters 8 to 10) were more in-depth and involved on-site visits, including stakeholder discussions and key informant interviews.

Table 1.3. Levels of HIA (adapted from Abrahams *et al.* 2004)

Level of HIA	Characteristics
Desk-based (or mini)	• Provides a broad overview of possible health impacts. • Collection and analysis of existing and readily accessible data.
Rapid (or standard)	• Provides more detailed information of possible health impacts. • Involves collecting and analysing existing data and some new qualitative data from stakeholders and key informants.
In-depth (or maxi)	• Provides a comprehensive assessment of potential health impacts. • Robust definition of impacts. • Involves collecting and analysing data using multiple methods and sources (quantitative and qualitative, including participatory approaches involving stakeholders and key informants). • Lasts six months or more.

1.7 CHALLENGES

There are a number of challenges associated with conducting HIA, some of these relate to the choice of methodology, e.g. quantification and stakeholder participation, while others apply to most assessments. These challenges are introduced below and some are re-examined, in light of the intervening chapters, in Chapter 11.

1.7.1 Stakeholder participation

The inclusion of stakeholder participation (and the level of that participation) will, largely depend upon the HIA method chosen and the level of HIA undertaken. Some have argued that stakeholder participation is an essential feature of HIA (Douglas *et al.* 2001), while others have suggested that it can be helpful in some contexts but not in others (Parry and Wright 2003). A range of levels of stakeholder participation is illustrated in the subsequent chapters.

The potential level of community participation in HIA is contextual. It depends on the prevailing governance system, on the constitutional framework, on local cultural practice, on determinants of the capacity of communities to participate (such as literacy rates) and on capacity among professional to apply participatory methods. For the institutional stakeholder participation, incentives and sanctions may be required, as intersectoral collaboration is not likely to materialize spontaneously. In most developing country contexts, if an adequate role for stakeholders is not spelled out in the CA, it is unlikely that they can be effective partners as the project evolves.

The rapid rural appraisal approach (Chambers 1983) provides a good framework for HIA stakeholder participation, as it recognizes the need to respect local power structures in formal contacts with communities, but also encourages researchers to sample, informally, views and opinions of community members who may have been marginalized.

1.7.2 Quantification

Quantification is, according to Veerman *et al.* (2005), comparatively rare in HIA. In a review of 98 prospective HIA studies only 17 reported quantitative estimates of exposure to determinants (e.g. environmental pollutants), of which 16 reported quantified health outcomes. Quantification can help decision-makers (Kemm 2000; Veerman *et al.* 2005) and allow the easy comparison of alternatives. It is clear, however, that not everything that is important can be quantified and sometimes it could be argued that "*not everything that can be quantified is important*" (Mindell *et al.* 2001). The main difficulties in terms of quantification are the availability of appropriate data and comparative metrics. Chapter 2 outlines a quantification method, using disability adjusted life years (DALYs), the use of which is illustrated in Chapters 3 to 7. The presentation of non-quantitative information is also explored in Chapter 2 and illustrated (using a variety of approaches) throughout subsequent Chapters.

1.7.3 Data needs and availability

Not surprisingly, one of the greatest challenges relates to the data needs for HIA and the availability of the data, this is especially true where quantification is an objective and when HIAs are to be done for projects, programmes and policies to be implemented in a developing country setting. Where desk-based or rapid HIAs are being conducted, there is little or no opportunity to collect new data, meaning that the best must be made of what is available. This often means resorting to routine surveillance data. It should be noted, however, that "*evidence for actual or potential impacts on health can come from many sources and a good HIA will strive to ensure different types of evidence are properly identified and considered*" (HDA 2002). As noted by Mindell *et al.* (2004) literature reviews may unearth conflicting evidence for health impacts or find that the evidence is either poor quality, insufficient or absent. Throughout this book, efforts have been made to summarize the best available evidence and clearly note where further data are required.

With more in-depth HIAs, such as that illustrated in Chapter 9, there may be scope for collecting data but resource is rarely available to facilitate the ideal level of data acquisition.

1.7.4 Possible biases

A main challenge in HIA is to prevent biases from creeping into the procedure and method. A quality HIA is objective, comprehensive and based on evidence. Biases can be introduced into the actual HIA procedure and method, and they can also be introduced at the time when recommendations are formulated outlining mitigation measures.

The terms of reference provide the first possible opportunity for biases, these are the product of the screening and scoping exercises, and false assumptions, poor judgement and wrong information can result in terms of reference that provide inadequate or outright wrong guidance to the HIA consultants, which will jeopardize the process from the outset.

A number of additional factors may contribute to a lack of objectivity within the HIA, including:

- conflicts of interest (between the client and consultant);
- inappropriate timing and/or duration;
- inadequate budgeting;
- lack of expertise, experience or skill; and
- lack of access to information.

1.7.5 Evaluation

Evaluation can take place at different stages of an HIA. Hopefully investigators evaluate the evidence used in their HIAs, however, the overall evaluation of HIA case studies in the literature is relatively rare (Taylor and Quigley 2002), although this may occur, behind closed doors, at the appraisal stage outlined in section 1.6. Clearly, to determine the effectiveness of HIAs it would be useful to monitor the process, the impact and the health outcomes (Taylor and Quigley 2002); i.e. how the process was undertaken, whether its recommendations were heeded and the health effects on the local population. Unfortunately, finances are rarely available for such follow-up work, especially in less developed countries where routine disease surveillance may be less than ideal (as illustrated in Chapters 8 and 9).

1.7.6 Multi-disciplinary nature

HIA is, by necessity, a multi-disciplinary approach, as the majority of proposals under investigation and many of the determinants of health which they affect are likely to be outside the field of public health (Mindell *et al.* 2004; Cole *et al.* 2005).

1.8 STRUCTURE OF THE BOOK

Chapter 2 sets the scene for the United Kingdom-based HIA research; it introduces the quantification methodology used and also a strategy and ranking system for using qualitative data. It outlines reference levels of risk used elsewhere and derives a screening level of risk, with figures below that level considered to be of no concern.

Chapter 3 is the first of the United Kingdom, desk-based case studies. It looks at using in-house rainwater harvesting for non potable uses, including toilet flushing. Chapters 4 and 5 examine the use of sustainable drainage options and greywater reuse, respectively, on the same population. Chapters 6 and 7 evaluate the health impacts of flooding in the United Kingdom, using literature-derived data, with Chapter 6 presenting an overall evaluation, while Chapter 7 focuses on a quantitative microbial risk assessment of flooding and gastrointestinal illness.

Chapters 8, 9 and 10 examine HIAs conducted in developing countries. Chapter 8 presents an early example, which was done to examine the possible impacts of a relatively small-scale irrigation scheme in Zimbabwe. Chapter 9 looks at a major hydroelectric project in Lao PDR. This HIA uses a visual risk profiling system, which outlines changes in risk to different vulnerable groups at different stages of the project, by environmental health area. Chapter 10 examines the health impacts of using raw wastewater in Pakistan to water food crops and weighs up the risks (such as infection) against the benefits (such as improved diet).

Chapter 11 presents a number of concluding remarks.

1.9 REFERENCES

Abrahams, D., Pennington, A., Scott-Samuel, A., Doyle, C., Metcalfe, O., den Broeder, L., Haigh, F., Mekel, O. and Fehr, R. (2004) European Health Impact Assessment – A Guide. ISBN 1 87403875 9.

Amerasinghe, F.P. (2003) Irrigation and mosquito-borne diseases. *Journal of Parasitology* **89** (Suppl.), S14–S22.

Arquiaga, M.C., Canter, L.W. and Nelson, D.I. (1994) Integration of health impact considerations in environmental impact studies. *Impact Assessment* **12**, 175–197.

Bangkok Charter (2006) Bangkok Charter on health promotion in a globalized world. www.who.int/healthpromotion/conferences/6gchp/hpr_050829_%20BCHP.pdf

Bell, H. (1999) *Frontiers of Medicine in the Anglo-Egyptian Sudan, 1899–1940.* Oxford Historical Monographs, Oxford University Press, Oxford.

Bond, A., Fawell, J., Harrison, R., Kay, D., Kemm, J., Kibble, A., Matthews, I., Mullins, J., Morgan, H., Palmer, S., Parry, J., Stoner, J. and Thomas, H. (2002) A health impact assessment of the landfill directive and government waste strategy. Environment Agency R&D Publication P6-011/1.

Bina, O. (2007) A critical review of the dominant lines of argumentation on the need for strategic environmental assessment. *Environmental Impact Assessment Review* **27**, 585–606.

Birley, M.H. (1991) Forecasting the vector-borne disease implications of water resources development. PEEM Guidelines series 2, WHO/CWS/91.3, World Health Organization, Geneva

Birley, M.H. (1995) *The Health Impact Assessment of Development Projects.* HMSO, London.

BMA (1998) *Health and Environmental Impact Assessment. An Integrated Approach.* British Medical Association. Earthscan Publications Ltd., London.

Bos, R., Birley, M.H., Furu, P. and Engel, Ch. (2003) Health opportunities in development – A course manual on developing intersectoral decision-making skills in support of health impact assessment.World Health Organization, Geneva

Bos, R. (2006) Health impact assessment and health promotion. *Bulletin of the World Health Organization* **84**, 914–915.

Chambers, R. (1983) *Rural Development: Putting the Last First.* Longman, Harlow.

Cole, B.L., Shimkhada, R., Fielding, J.E., Kominski, G. and Morgenstern, H. (2005) Methodologies for realizing the potential of health impact assessment. *American Journal of Preventive Medicine* **28**, 382–389.

deFur, P.L., Evans, G.W., Cohen Hubal, E.A., Kyle, A.D., Morello-Frosch, R.A. and Williams D.R. (2007) Vulnerability as a function of individual and group resources in cumulative risk assessment. *Environmental Health Perspectives* **115**, 817–824.

DoH (1995) The Health of a Nation: Policy Appraisal and Health; A Guide from the Department of Health. Department of Health, London.

Douglas, M., Conway, L., Gorman, D., Gavin, S. and Hanlon, P. (2001) Developing principles for health impact assessment. *Journal of Public Health Medicine* **23**, 148–154.

enHealth (2001) Health Impact Assessment Guidelines. Commonwealth of Australia.

Erlanger, T.E., Krieger, G.R., Singer, B.H. and Utzinger, J (2008) The 6/94 gap in health impact assessment. *Environmental Impact Assessment Review* **28**, 349–358.

Ewan, C., Young, A., Bryant, E. and Calvert, D. (1994) National Framework for Environmental and Health Impact Assessment. NHMRC, Australia.

Fewtrell, L., Kay, D., Dunlop, J., O'Neill, G. and Wyer, M. (1994) Infectious diseases and water-supply disconnections. *Lancet* **343**, 1370.

Fewtrell, L. and Kay, D. (2008) Health impact assessment of water management options: generic guidance note. UK Water Industry Research, London.

Goodland, R. (1995) The concept of environmental sustainability. *Annual Review of Ecology and Systematics* **26**, 1–24.

Goodland, R.J.A. (2000) New methodologies and instruments in sustainable river basin management. Is there a future for big dams? In *New Approaches to River*

Management (eds A.J.M. Smith, P.H. Nienhuis and R.S.E. Leuven), pp. 187–207, Backhuys Publishers, Leiden.

HDA (2002) Introducing health impact assessment (HIA): informing the decision-making process. Health Development Agency, London.

IAIA/WHO (2006) Health impact assessment: international best practice principals. http://www.iaia.org/modx/assets/files/SP5.pdf accessed 14 April 2008.

IAIA (2007) http://www.iaia.org/modx/index.php?id = 318 accessed 14 April 2008.

IPIECA (2005) A guide to health impact assessments in the oil and gas industry. International Petroleum Industry Environmental Conservation Association. http://www.ipieca.org/activities/health/downloads/publications/hia.pdf accessed 14 April 2008.

Jobin, W.R. (1999) *Dams and Disease. Ecological Design and Health Impacts of Large Dams, Canals and Irrigation Systems.* E & F N Spon, London and New York.

Jobin, W. (2003) Health and equity impacts of a large oil project in Africa. *Bulletin of the World Health Organization* **81**, 420–426.

Joffe, M. and Mindell, J. (2002) A framework for the evidence base to support health impact assessment. *Journal of Epidemiology and Community Health* **56**, 132–138.

Kemm, J.R. (2000) Can health impact assessment fulfil the expectations it raises? *Public Health* **114**, 431–433.

Kemm, J. and Parry, J. (2004) What is HIA? Introduction and overview. In *Health Impact Assessment: Concepts, Theory, Techniques, and Applications* (eds J. Kemm, J. Parry and S. Palmer) pp. 1–13, Oxford University Press Inc., New York.

Kemm, J., Parry, J. and Palmer S. (2004) *Health Impact Assessment: Concepts, Theory, Techniques, and Applications.* Oxford University Press Inc., New York, pp. 432.

Krieger, N., Northridge, M., Griskin, S., Quinn, M., Kriebel, D., Davey Smith, G., Bassett, M., Rehkopf, D.H., Miller, C. and the HIA "promise and pitfalls" conference group (2003) Assessing health impact assessment: multidisciplinary and international perspectives. *Journal of Epidemiology and Community Health* **57**, 659–662.

Konradsen, F., Chimbari, M.J., Birley, M., Furu, P. and Christensen, N. (1997) The use of health impact assessments in water resources development: a case study from Zimbabwe. *Impact Assessment* **1**, 55–73.

Lock, K. (2000) Health impact assessment. *British Medical Journal* **320**, 1395–1398.

Mindell, J., Boaz, A., Joffe, M., Curtis, S. and Birley, M. (2004) Enhancing the evidence base for health impact assessment. *Journal of Epidemiology and Community Health* **58**, 546–551.

Mindell, J., Hansell, A., Morrison, D., Douglas, M. and Joffe, M. (2001) What do we need for robust, quantitative health impact assessment? *Journal of Public Health Medicine* **23**, 173–178.

Moe, C.L. and Rheingans, R.D. (2006) Global challenges in water, sanitation and health. *Journal of Water and Health* **4** (Suppl 1), 41–57.

Montgomery, E., Bennett, J. W. and Scudder, T. (1973) The impact of human activities on the physical and social environments: new directions in anthropological ecology. *Annual Reviews of Anthropology* **2**, 27–61.

National Health Service Executive (2001) Health impact assessment. Volumes 1 and 2. National Health Service, London.

NHMRC (1994) National framework for environmental and health impact assessment. National Health and Medical Research Council. Commonwealth of Australia.

NYPHO (2001) An overview of health impact assessment. Occasional Paper no. 1. Northern and Yorkshire Public Health Observatory.

Obrist, B., Iteba, N., Lengeler, C., Makemba, A., Mshana, C., Nathan, R., Alba, S., Dillip, A., Hetzel, M.W., Mayumana, I., Schulze, A. and Mshinda, H. (2007) Access to health care in contexts of livelihood insecurity: a framework for analysis and action. *PLoS Medicine* **4**, 1584–1588.

Ottawa Charter (1986) Ottawa Charter on Health Promotion http://www.who.int/hpr/NPH/docs/**ottawa_charter**_hp.pdf accessed April 14 2008.

Parry, J. and Kemm, J. (2004) Future directions for HIA. In *Health Impact Assessment: Concepts, Theory, Techniques, and Applications* (eds J. Kemm, J. Parry and S. Palmer), pp. 411–417, Oxford University Press Inc., New York.

Parry, J. and Wright, J. (2003) Community participation in health impact assessments: intuitively appealing but practically difficult. *Bulletin of the World Health Organization* **81**, 388.

Petticrew, M., Macintyre, S. and Thomson, H. (2004) Evidence and HIA? In *Health impact assessment: concepts, theory, techniques, and applications* (eds J. Kemm, J. Parry and S. Palmer), pp. 71–80, Oxford University Press Inc., New York.

PHRB (1994) Health impact assessment toolkit. Population Health Resource Branch, Ministry of Health, Vancouver, British Columbia, Canada.

Putters, K. (1994) Health impact screening. Ministry of Health, Welfare and Sport, Rijswijk, The Netherlands.

Rosegrant, M.W. and Cline, S.A. (2003) Global food security: challenges and policies. *Science* **302**, 1917–1919.

Scudder, T. (2005) *The Future of Large Dams*. Earthscan, London.

Sir M. MacDonald and Partners, Hunting Technical Services Ltd. and East African Engineering Consultants (1977) Bura Irrigation Settlement Project, Project Planning Report volume 4. Republic of Kenya, Ministry of Agriculture, National Irrigation Board, Kenya.

Smith, A.H., Lingas, E.O. and Rahman, M. (2000) Contamination of drinking-water by arsenic in Bangladesh: a public health emergency. *Bulletin of the World Health Organization* **78**, 1093–1103.

Steinemann, A. (2000) Rethinking human health impact assessment. *Environmental Impact Assessment Review* **20**, 627–645.

Steinmann, P., Keiser, J., Bos, R., Tanner, M. and Utzinger, J. (2006) Schistosomiasis and water resources development: systematic review, meta-analysis, and estimates of people at risk. *Lancet Infectious Diseases* **6**, 411–425.

Southgate, V.R. (1997) Schistosomiasis in the Senegal River Basin: before and after the construction of the dams at Diama, Senegal and Manantali, Mali and future prospects. *Journal of Helminthology* **71**, 125–132.

Taylor, L. and Quigley, R. (2002) Health impact assessment: a review of reviews. National Health Service, Health Development Agency, London.

Tiffen, M. (1991) Guidelines for the incorporation of health safeguards into irrigation projects through intersectoral cooperation with special reference to vector-borne diseases. PEEM Guidelines Series 1. WHO/CWS/91.2. World Health Organization, Geneva.

Utzinger, J. (2004) Book review: Health impact assessment: concepts, theory, techniques, and applications. *Bulletin of the World Health Organization* **82**, 954.

Utzinger, J., Wyss, K., Moto, D.D., N'Diekhor, Y., Tanner, M. and Singer, B.H. (2005) Assessing health impacts of the Chad-Cameroon petroleum development and pipeline project: challenges and a way forward. *Environmental Impact Assessment Review* **25**, 63–93.

Veerman, J.L., Barendregt, J.J. and Mackenbach, J.P. (2005) Quantitative health impact assessment: current practice and future directions. *Journal of Epidemiology and Community Health* **59**, 361–370.

Vu, S.N., Nguyen, T.Y., Kay, B.H., Marten, G.G. and Reid, J.W. (1998) Eradication of *Aedes aegypti* from a village in Vietnam, using copepods and community participation. *American Journal of Tropical Medicine and Hygiene* **59**, 657–660.

WCD (2000) *Dams and Development: a new framework for decision-making*. The Report of the World Commission on Dams. Earthscan Publications Ltd., London.

WHO (1946) Preamble to the Constitution of the World Health Organization as adopted by the International Health Conference, New York, 19 June – 22 July 1946.

WHO (1983) Report of the third meeting of the joint WHO/FAO/UNEP Panel of Experts on Environmental Management for Vector Control (PEEM). Part II: Technical Discussion – methods of forecasting the vector-borne disease implications in the development of different types of water resources projects. Document VBC/83.4, World Health Organization, Geneva.

WHO (2002) Technical briefing: Health impact assessment – a tool to include health on the agenda of other sectors. Current experience and emerging issues in the European Region. WHO Regional Committee for Europe, Copenhagen.

WHO (2007) Water scarcity fact file. World Health Organization, Geneva. http://www.who.int/features/factfiles/water/water_facts/en/index.html accessed April 14 2008.

WHO ECHP (1999) Health Impact Assessment: main concepts and suggested approach. Gothenburg consensus paper. World Health Organization European Centre for Health Policy, Brussels, Belgium.

2

An overview of UK-based HIA research on water management and flooding

Lorna Fewtrell, David Butler, Fayyaz Ali Memon, Richard Ashley and Adrian Saul

2.1 INTRODUCTION

This part of the book has been derived from a series of UK-based research projects funded by the Engineering and Physical Sciences Research Council (EPSRC), each of which contained components examining the health impacts of certain water-related issues. This Chapter serves as an introduction to these projects and the subsequent Chapters (3 to 7) within this part, and outlines material common to the approach taken.

2.1.1 Water Cycle Management for New Developments (WaND)

In response to a rising population, changes in demographic structure, economic migration across the European Union and a lack of affordable housing, the UK government has embarked on an ambitious house building programme. The housing pressures, and hence the bulk of the proposed housing growth, are most acute in the south-east of England, which is now increasingly acknowledged as a water stressed region. Delivering water-related services to these new developments requires a shift in the traditional approach from managing the three principal urban water flows (i.e. potable water, stormwater and wastewater) as individual entities towards exploring and exploiting recent innovations and managing the system as an integrated whole. As a contribution to this need, the Engineering and Physical Science Research Council, the Environment Agency and a number of water service providers funded a 4 – year research project led by the University of Exeter on water cycle management for new developments (WaND). WaND was a multi-disciplinary and multi-institutional research team addressing sustainable water management technology (for example, water saving devices, greywater reuse, rainwater harvesting, SUDS and small bore wastewater collection systems) including its social, health and environmental implications, innovation uptake and risk mitigation strategies, sustainability, spatial planning, decision support and decision making. The primary aim was to support the delivery of integrated, sustainable water management for new developments by provision of tools and guidelines for project design, implementation and management. The project's outputs are available on the project web-site (www.wand.uk.net) and include a stakeholder specific knowledge base and decision support mechanism for the planning and implementation of sustainable solutions.

2.1.2 Adaptable Urban Drainage – Addressing Change in Intensity, Occurrence and Uncertainty of Stormwater (AUDACIOUS)

The main aim of AUDACIOUS was to investigate key aspects of climate change and urbanisation on existing drainage in local urban areas and provide tools for a range of stakeholders to adapt to uncertain future scenarios. The focus was on adapting to flooding caused by a lack of capacity or hindrance (e.g., blockage) of the urban drainage system, rather than the inundation of buildings on river flood plains due to over-bank flows from major watercourses (Ashley *et al.* 2007).

Although aimed at existing systems, AUDACIOUS also considered new build and set out to link with the other primary areas – interaction of urban floods with other urban infrastructure (e.g., transportation, communications and buildings) and to tackle some of the many cross-cutting issues (e.g. socio-economic and health implications, stakeholder perception, planning, insurance risk and land development).

2.1.3 Flood Risk Management Research Consortium (FRMRC)

In 2003, EPSRC, in collaboration with the Department for Environment, Food and Rural Affairs/Environment Agency (DEFRA/EA) joint research and development programme on Flood and Coastal Defence, the Natural Environment Research Council (NERC), the Scottish Executive, the Rivers Agency (Northern Ireland) and UK Water Industry Research (UKWIR) funded an interdisciplinary research consortium, the Flood Risk Management Research Consortium, to improve understanding in the prediction and management of flood risk. The research portfolio was co-ordinated under eight research priority areas and included the short-term delivery of tools and techniques to support more accurate flood forecasting and warning, improvements to flood management infrastructure and a reduction of flood risk to people, property and the environment through the development of sustainable flood management strategies.

2.1.4 Health impact assessment

The health impact assessments outlined in Chapters 3 to 7, concentrate on the assessment step outlined in Chapter 1 (focussing on risk assessment, risk management and the health impact statement). As they are all literature/desk-based studies there is no component of empirical data acquisition and no option for post-project monitoring and evaluation. In order to avoid repetition, much of the material common to the HIAs is outlined in this Chapter.

2.2 QUANTIFICATION

The HIAs outlined in Chapters 3 to 7 seek to quantify risk wherever possible. The chosen quantification metric was the Disability Adjusted Life Year (DALY), which allows the comparison between different scenarios and produces an objective measure of overall risk (see section 2.2.1). It is, however, fully acknowledged that not everything can be quantified with numerical metrics and, to address this, a qualitative system was also devised to provide guidance to planners, developers and users.

2.2.1 DALYs

DALYs are summary measures of health that allow the comparison of effects across a wide range of health outcomes, including mortality and morbidity. The measure combines years of life lost by premature mortality (YLL) based, for estimations of global disease burden, on a standard life expectancy at birth of 80.0 years for men and 82.5 years for women (Murray 1996), with years (or days, weeks or months) lived with a disability (YLD), standardized using severity or disability weights. The weights range from 0 (perfect health) to 1 (dead). Thus a death from drowning would have a severity score of 1. If drowning occurred in a child aged 2.5 the DALY score would be 76.5 (based on the average UK life expectancy of 79 [GAD 2007] minus 2.5). This can be compared with a mild case of gastrointestinal illness (stomach upset), where the severity weight is 0.093 and the duration is (on average) three days (YLL = severity * duration), resulting in a score of 7.6×10^{-4}.

Severity scores have been taken directly from the literature where possible (VGDHS 1999), or extrapolated from similar conditions where scores have not previously been derived. Examples from an illustrative range of outcomes (not necessarily related to water management issues) are shown in Table 2.1 (VGDHS 1999).

Table 2.1. Severity weights for a range of health outcomes

Disease category, subcategory or sequelae	Severity weight
Diarrhoeal diseases and gastroenteritis	
Uncomplicated episode	0.093
Complicated episode	0.420
Acute respiratory infections	
Influenza episode	0.047
Acute bronchitis episode	0.132
Pneumonia episode	0.373
Fractures	
Pelvis	0.247
Radius or ulna	0.180
Ankle	0.196

In some applications (such as the Global Burden of Disease study – Murray 1996), DALYs are subject to weighting for age and discounting of loss of healthy life in the future. Neither of these features is central to the concept of

DALYs (Havelaar and Melse 2003) and no weighting or discounting has been applied in the research reported here.

2.2.2 Qualitative data

For a variety of reasons, some things can not be quantified but may be important in terms of health impacts (e.g. microbial infections where no dose-response relationship has been established, the impact of a new/unfamiliar technique, such as greywater reuse, on anxiety levels). These have been considered on a case by case basis, in a similar way to the impacts that can be quantified. Areas examined include:

- the current level (incidence) of the problem in the UK, noting where there are uneven distributions (e.g., in relation to age group);
- whether the issue under consideration (e.g., the water management option) could lead to an increase, or decrease, in the health outcome and, if so, by what mechanism; and
- the likely consequences (e.g., death, serious injury or illness, mild illness).

Based on these considerations, impacts have been given a qualitative ranking (ranging from − − − to + + +) to indicate the possible magnitude of the problem. Thus, an estimate of − − − suggests a high negative impact in terms of both increased exposure to the cause and a severe outcome, while (−) indicates a possible but unlikely (or minor) negative health impact, while positive estimates suggest a health benefit.

2.3 SCREENING

In Chapters 3 to 5, health impact assessments are conducted on a hypothetical case study area, based in the South of England (see section 2.5). A series of water management options were examined and the possible health impacts of these options explored.

2.3.1 Water management options

The water management options and their application to the case study development are summarized below:

- **Rainwater harvesting**. Each property was assumed to have an in-house rainwater harvesting system, with an underground tank, collecting rainwater from the roof via downpipes, passing through a filter on entry to the tank. The tank receives top up supplies from the mains water when it is running low and the supplies are used for toilet flushing and garden watering. The system is the responsibility of the individual householder.

- **Sustainable drainage**. The sustainable drainage system (SUDS) assessed consisted of a pre-existing wetland, a retention pond, a number of detention basins (which often contained small amounts of water) and six swales (although these could be described as linear ponds, as they are more like long, steep-sided ditches).
- **Greywater reuse**. Properties were assumed to have a vertical flow constructed wetland (reedbed) and chlorine disinfection at household-level. The system takes water from showers, baths and wash basins and the treated water is used for external purposes (e.g. garden watering and car washing).

The water management options were chosen following consultation with the other project investigators. Some of the forms in which the options were applied (notably the greywater reuse) were done for exploratory purposes, rather than as a realistic application of the technology. In each case, it was considered that possible health impacts could be significant and, as no health impact assessments had previously been conducted looking at these areas, potentially useful information was likely to result from the HIA examination.

2.3.2 Flood risk

In Chapters 6 and 7, flood risk on two separate case study populations is investigated. The characteristics of the flooding and the populations affected are outlined within their respective Chapters.

2.4 SCOPING

Scoping lays the foundations for the remaining HIA steps. The key task in the scoping process is setting the boundaries for the assessment. This is illustrated below in relation to the WaND project HIAs:

- **Timescale**. This was defined as a year from when the housing development is complete and (essentially) fully occupied.
- **Geographic boundaries**. These are restricted to the boundary of the housing development. No consideration was given to the possible impact of the site and the water management options on other locations.

- **Population covered**. See section 2.5.
- **Vulnerable, disadvantages or marginalised groups**. See section 2.5.
- **Potential health effects/impacts of concern**. It was considered that there could be effects from some of the water management options on the risk of injury, communicable disease (i.e. infection), anxiety and physical activity to the resident population.

2.5 PROFILING (CASE STUDY)

The case study population for the water management option HIAs (Chapters 3 to 5) was loosely based on the population of a newly built housing development in the South of England comprising over 1800 houses, with the age structure shown in Table 2.2. Almost 90% of the population classified their health as good (compared to 69% of the general UK population).

Table 2.2. Case study socio-demographic information

	Case study area		England
Category	Number	Percentage	Percentage
General			
Population density	2.54		2.4
Number of houses	1868		
Total population	4746	100	
Age groups			
0–4	471	9.9	6.0
5–15	634	13.4	14.2
16–64	3532	74.4	63.9
65+	109	2.3	15.9

2.5.1 Vulnerable groups

There are a number of factors that make some people more vulnerable to health impacts, as illustrated in Table 2.3 (adapted from Hunter and Fewtrell 2001). In the HIAs reported in this part of the book, a number of vulnerability factors have been considered, including age (which partly determines differential vulnerabilities to infection and drowning), pre-existing disease (by accounting for the level of HIV/AIDS in the community and the likely impact this will have on infection) and behaviour.

Table 2.3. Selected factors leading to inequality of health risk

Factor	Affects
Age	The very young and very old are more likely to acquire infections due to naïve or waning immunity and, once infected, are more likely to develop severe outcomes. The elderly may be more anxious about unfamiliar systems or maintenance requirements.
Pre-existing disease	A person with AIDS or severe combined immunodeficiency syndrome is likely to suffer far more severe symptoms with cryptosporidiosis and some other infectious diseases
Genetic	People with certain genotypes are more likely to experience complications, such as joint problems, following gastro-intestinal infections
Gender/pregnancy	Certain infections are more severe in pregnancy, either increasing the risk of fatality for the woman, or damage to the foetus
Behaviour	Behaviour patterns, such as geophagy (soil eating), may result in higher exposure to infectious and toxic agents

2.6 RISK ASSESSMENT

Classical risk assessment consists of four stages (in which the first three steps are combined in order to characterize the expected risk), namely:

- hazard identification (or characterization);
- exposure assessment;
- dose-response assessment (applicable to microbial and chemical hazards); and
- risk characterization.

For certain microbial risks quantitative microbial risk assessment (QMRA) techniques have been applied as part of the HIA. QMRA is a formal probabilistic approach for estimating microbial risks associated with defined scenarios. In a number of cases, parameters have been represented as probability distributions, rather than point estimates, in order to examine the effects of uncertainty. In the studies reported here Monte Carlo sampling (5000 iterations) was used for each simulation run using @Risk version 4.5 Professional edition (Palisade Corporation 2002), this resulted in minimum, maximum and mean estimates of DALYs, with the mean value being used in the health impact statement.

2.6.1 Hazard identification

A health hazard is anything that can potentially cause harm; with harm being loss of life, injury, illness and so on. A number of hazards are common across many of the water-related HIAs outlined in this book, including;

- drowning and near-drowning;
- unintentional injury;
- infection; and
- anxiety.

Some of these hazards (such as drowning and near-drowning) are, perhaps, more obvious than others (such as falls relating to maintenance activities).

2.6.2 Exposure assessment

The techniques to estimate exposure vary according to the specific health impact and may require a number of assumptions to be made. Where possible these have been based on the literature and clearly outlined.

2.6.3 Dose-response assessment

For chemical and microbial hazards a dose-response assessment is necessary, as exposure to the hazard does not necessarily mean that a health impact is inevitable (Figure 2.1), and the dose-response assessment characterizes the relationship between exposure and the incidence of the health effect(s). In many cases, depending upon what dose is actually received (and how it is received), the body may be able to remove the toxin or pathogenic micro-organism without any ill effect. In drinking-water, for example, numerous chemicals may be present (cadmium, for example), which, providing they are below specified

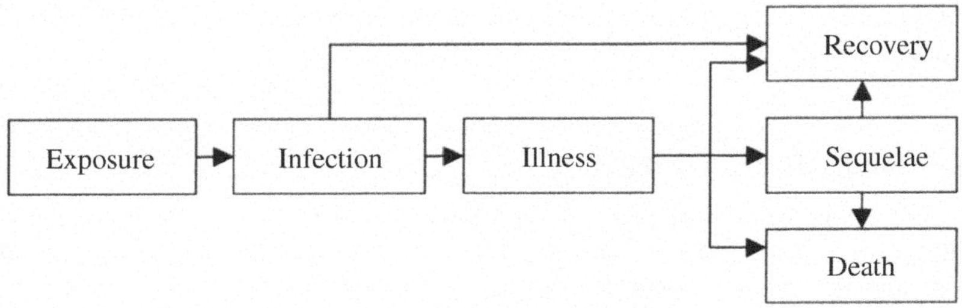

Figure 2.1. The infectious disease process (FAO/WHO 2003)

levels (i.e. 0.003 mg/litre in the case of cadmium), present a negligible hazard to health in the normal population (WHO 2004).

Few chemicals were considered to pose possible hazards in terms of the water management options considered, so the discussion here is confined to microbial dose-response relationships. The response of individuals to various exposures of pathogenic microorganisms is a particularly grey area; not least because of ethical considerations. A number of experiments in humans to determine dose-response relationships for a variety of pathogens have been performed, which provide valuable information, but these tend to be conducted at relatively high doses (as progressively larger numbers of people are required as the dose decreases). Thus, even where data exist they generally have to be extrapolated to cover the low doses that people would be expected to encounter in normal circumstances. In addition, in should be noted that experimental infection for generation of dose-response information is usually done in relatively young and healthy adults (a population group which may not be representative of the whole population and in particular any susceptible sub-population fractions).

A number of researchers (most notably Chuck Haas) has evaluated the applicability of a variety of mathematical models in describing the dose-response data (see for example Haas 1983; Haas *et al.* 1993; Haas *et al.* 1997; Holcomb *et al.* 1999). Haas (1983) found that the beta-Poisson model appeared to be the most widely applicable, but noted that the models examined gave divergent predictions when used to extrapolate to very low exposure levels. It is now accepted (Haas and Eisenberg 2001) that dose-response data may be adequately described by either the exponential or beta-Poisson model (see Box 2.1).

Table 2.4 shows the parameters of a number of dose-response relationships that have been used in QMRA in order to estimate infection.

As infection precedes illness (and in some cases death), as illustrated in Figure 2.1, the probability of infection is usually treated as the worst case estimate for the least risk of illness (Teunis *et al.* 1994) and is considered to be conservative. In most cases, however, the generalized infection to illness rates are reported in the peer-reviewed literature allowing clinical illness to be estimated.

2.6.4 Risk characterization

The risk characterization process combines the information from the hazard identification, exposure assessment and, if appropriate, the dose-response assessment to derive an overall assessment for each hazard.

Exponential model
This is derived from the assumption of random occurrence of microorganisms along with a constant probability of initiation of infection by a single organism r, the probability of infection P_I is given as a function of the ingested dose d by:

$$P_I = 1 - \exp^{(-rd)} \quad (1)$$

Beta-Poisson model
For many microorganisms, the dose-response relationship is shallower than reflected by equation 1, suggesting some degree of heterogeneity in the microorganism-host interaction. This can be successfully described by the beta-Poisson model, which can be developed from equation 1 if the infection probability is itself distributed according to a beta distribution. This model is described by two parameters, a median infectious dose N_{50} and a slope parameter α:

$$P_I = 1 - [1 + d/N_{50}\,(2^{1/\alpha} - 1)]^{-\alpha} \quad (2)$$

This can be written more simply (McNab 1997) where

$$\beta = N_{50} / (2^{1/\alpha} - 1) \quad (3)$$

thus: $P_I = 1 - (1 + d/\beta)^{-\alpha}$ (4)

Given a set of dose-response data, i.e. exposure of population to various doses of microorganisms and measurement of response (such as infection), the best fitting parameters of a dose-response relationship may be computed via maximum likelihood techniques.

Box 2.1. Dose-response models

Table 2.4. Selected dose-response parameters

	Exponential	Beta-Poisson		
Organism	r	α	β	Reference
Cryptosporidium parvum	0.004005			Teunis *et al.* 1996; DuPont *et al.* 1995
Campylobacter jejuni		0.145	7.589	Medema *et al.* 1996; Black *et al.* 1988
Salmonella spp.		0.313	2884	Haas *et al.* 1999
S. meleagridis		0.89	4.4×10^5	Teunis *et al.* 1999

2.7 HEALTH IMPACT STATEMENT AND REFERENCE LEVELS OF RISK

The characterized risks for each of the identified hazards are brought together to derive a health impact statement, with the results presented on the basis of overall scores for both DALYs and qualitative information. The use of DALYS in health impact assessment is at an early stage and so there is little information to guide the assessor. The main guidance, in terms of DALYs, that has been located in the literature relates to a reference level of risk of 10^{-6} DALYS per person per year for drinking-water (which is approximately equivalent to a lifetime excess cancer risk of 10^{-5}) suggested by WHO (2004); a risk level of 10^{-6} DALYS/person/year has also been adopted as a health-based target in setting guidelines for the safe use of wastewater, excreta and greywater (e.g. WHO 2006).

Considering guidance from other areas, reference levels are often set according to the probability or odds of death. For example, the Health and Safety Executive (HSE) consider that an annual probability of death of 1 in 1,000,000 is a level of risk at which no further improvements in safety are justified. This level was decided upon by considering risks in other contexts; with odds of 1 in 1,000,000 being roughly the same as an electrocution incident at home and one hundredth that of dying in a road traffic accident (RCEP 1998). Assuming that death is most likely in the middle years, say 40, an annual fatality level of 1 in a million would translate to a DALY score of 3.9×10^{-5} (life expectancy of 79 – age at death 40 = 39/1,000,000). This reference level can be compared with selected annual fatalities in the UK from 2005 (National Statistics 2006) converted into DALYs (Table 2.5).

Table 2.5. Selected annual mortality, equivalent DALY scores and reference risk levels

Cause	No of deaths in 2005	Total YLL*	Equivalent DALYS per person
Pedestrian killed in transport accident	543	17,498	3.2×10^{-4}
Drowning in the bath	30	1022	1.9×10^{-5}
Electrocution	28	908	1.7×10^{-5}
Lightening strike	2	111	2.1×10^{-6}
Reference level (HSE)	–	–	3.9×10^{-5}
Reference level (WHO)	–	–	1×10^{-6}

* YLL – years of life lost, based on the number of deaths in each age group (by gender) multiplied by the average life expectancy for that age and gender group.

Clearly, DALYS can also accommodate injury and illness. Thus, a situation might be considerable 'acceptable' in HSE terms if it results in death in no more than one case in a million, even though it results in a high burden of non-fatal injury or illness. It is, therefore, important to consider non-fatal outcomes in relation to an appropriate reference level.

In the USA, the concept of 10^{-6} (a one-in-a-million risk of developing cancer) has become enshrined as a level of acceptable risk, although it was originally developed as a level of risk below which was considered a 'trifle' and not of regulatory concern (Kelly 1991), i.e. essentially zero. However, instead of being seen as effectively zero, this level has often been interpreted as a maximum acceptable level of risk. Ultimately the definition of an 'acceptable' reference level of risk is political. However, for the purposes of research reported in this part of the book, an annual DALY score of 5×10^{-5} per person is suggested as a screening level, with figures below that considered to be of no concern.

2.8 DISCUSSION

Risk assessments of any kind are normally generalized and, while estimates may be reasonable at a group or population level, they are not expected to be accurate at an individual level and there will always be a disproportionate burden of disease (even where vulnerable groups are considered and accounted for). For example, three cases of campylobacteriosis in the case study population outlined in section 2.5 would give a DALY score of 2.1×10^{-6} per person. These will clearly not, however, be distributed throughout the whole population, but will individually affect three people.

In the health impact assessments outlined in Chapters 3 to 5, the water management options have been assumed to be correctly maintained and operating as expected. Poor maintenance and/or non-robust systems ***will*** change the level of risk and consequent health impact. Additionally, only the risks to the resident population have been considered; no consideration has been given to maintenance personal.

Systems deteriorate (even with timely and appropriate maintenance), people adapt and may modify their behaviour and the climate may change (potentially leading to emerging diseases and/or changing vectors of disease). These factors indicate that health impact assessment results are likely to need periodic revisiting and reassessment. A well conducted health impact assessment essentially predicts the effects of a system at a point in time, based on current knowledge and data.

2.9 REFERENCES

Ashley, R.M., Blanksby, J.R., Cashman, A., Jack, L., Wright, G., Packman, J., Fewtrell, L., Poole, A. and Maksimovic, C. (2007) Adaptable urban drainage – addressing change in intensity, occurrence and uncertainty in stormwater (AUDACIOUS). *Journal of the Built Environment* **33**(1), 70–84.

Black, R.E., Levine, M.M., Clements, M.L., Hughes, T.P. and Blaser, M.J. (1988) Experimental *Campylobacter jejuni* infection in humans. *The Journal of Infectious Diseases* **157**(3), 472–479.

DuPont, H.L., Chappell, C.L., Sterling, C.R., Okhuysen, P.C., Rose, J.B. and Jakubowski, W. (1995) The infectivity of *Cryptosporidium parvum* in healthy volunteers. *New England Journal of Medicine* **332**(13), 855–859.

FAO/WHO (2003) Hazard characterization for pathogens in food and water. Guidelines. Microbiological Risk Assessment Series, No 3. Food and Agriculture Organization of the United Nations/World Health Organization.

GAD (2007) Interim Life Tables 2003–2005. UK. Male and female. Government's Actuary Department, UK Government, London, UK.

Haas, C.N. (1983) Estimation of risk due to low doses of microorganisms: a comparison of alternative methodologies. *American Journal of Epidemiology* **118**(4), 573–582.

Haas, C.N., Rose, J.B., Gerba, C. and Regli, S. (1993) Risk assessment of virus in drinking water. *Risk Analysis* **13**, 545–552.

Haas, C.N., Rose, J.B., Gerba, C.P. and Crockett, C.S. (1997) What predictive food microbiology can learn from water microbiology. *Food Technology* **51**(4), 91–94.

Haas, C.N., Rose, J.B. and Gerba, C.P. (1999) *Quantitative Microbial Risk Assessment*. Wiley, New York.

Haas, C. and Eisenberg, J.N.S. (2001) Risk assessment. In *Water Quality: Guidelines, Standards and Health* (eds L. Fewtrell and J. Bartram), pp. 161–183, World Health Organization Water Series. IWA Publishing, London.

Havelaar, A.H. and Melse, J.M. (2003) Quantifying public health risk in the WHO Guidelines for Drinking-Water Quality. RIVM Report 734301022/2003, Bilthoven, The Netherlands.

Holcomb, D.L., Smith, M.A., Ware, G.O., Hung, Y.C., Brackett, R.E. and Doyle, M.P. (1999) Comparison of six dose-response models for use with food-borne pathogens. *Risk Analysis* **19**(6), 1091–1100.

Hunter, P.R. and Fewtrell, L. (2001) Acceptable risk. In *Water Quality: Guidelines, Standards and Health*. (eds L. Fewtrell and J. Bartram), pp. 207–227, World Health Organization Water Series. IWA Publishing, London.

Kelly, K.E. (1991) The myth of 10^{-6} as a definition of acceptable risk. Paper presented at the 84th annual meeting of the Air and Waste Management Association, Vancouver, BC, Canada. www.deltatoxicology.com/pdf/10–6.pdf (accessed 31/10/07).

McNab, W.B. (1997) A literature review linking microbial risk assessment, predictive microbiology and dose-response modelling. *Dairy, Food and Environmental Sanitation* **17**(7), 405–416.

Medema, G.J., Teunis, P.F.M., Havelaar, A.H. and Haas, C.N. (1996) Assessment of the dose-response relationship of *Campylobacter jejuni*. *International Journal of Food Microbiology* **30**, 101–111.

Murray, C.J.L. (1996) Rethinking DALYs. In *The Global Burden of Disease and Injury Series. Volume I. The Global Burden of Disease* (eds C.J.L. Murray and A.D. Lopez), pp. 1–98, Boston: Harvard School of Public Health, World Health Organization, World Bank.

National Statistics (2006) Mortality statistics: cause. Review of the Registrar General on deaths by cause, sex and age, in England and Wales, 2005. Series DH2 no. 32. Office for National Statistics, London.

Palisade Corporation (2002) Guide to using @Risk: risk analysis and simulation add in for Microsoft Excel. Version 4.5. Palisade Corporation, Newfield, New York. http://www.palisade.com accessed April 14 2008.

RCEP (1998) Setting Environmental Standards. Royal Commission on Environmental Pollution. 21st Report, The Stationery Office, London.

Teunis, P.F.M., Havelaar, A.H. and Medema, G.J. (1994) A literature survey on the assessment of microbiological risk for drinking water. Report 734301006. RIVM, Bilthoven, The Netherlands.

Teunis, P.F.M., van der Heijden, O.G., van der Giessen, J.W.B. and Havelaar, A.H. (1996) The dose-response relation in human volunteers for gastro-intestinal pathogens. RIVM Report 284550002, Bilthoven, Netherlands.

Teunis, P.F.M., Nagelkerke, N.J.D. and Haas, C.N. (1999) Dose response models for infectious gastroenteritis. *Risk Analysis* **19**(6), 1251–1260.

VGDHS (1999) Victorian Burden of Disease Study: Morbidity. Victorian Government Department of Human Services, Melbourne, Victoria, Australia.

WHO (2004) Guidelines for Drinking-water Quality. Third Edition. Volume 1 Recommendations. World Health Organization, Geneva.

WHO (2006) WHO Guidelines for the Safe Use of Wastewater, Excreta and Greywater. Volume II: Wastewater Use in Agriculture. World Health Organization, Geneva.

3

Rainwater harvesting – an HIA of rainwater harvesting in the UK

Lorna Fewtrell, David Kay and Adrian McDonald

3.1 INTRODUCTION

Rainwater harvesting is a very simple concept, where rain falling onto a roof surface is collected and subsequently used as a source of water. It has been in use in some areas (e.g. Israel, Africa and India) probably since 2000 BC and is now used in both developing and developed countries where it provides both potable (e.g. Australia, Canada, USA) and non-potable supplies (e.g. Germany).

As the predicted use of rainwater harvesting in the UK is for domestic non-potable use (i.e., toilet flushing, laundry use, garden watering etc.) most of the discussion here is related to non-potable supplies, although experience from potable systems is drawn upon in assessing possible health effects (through inadvertent or deliberate consumption of harvested rainwater).

The simplest rainwater collection method, and the one with the longest history in the UK, is the garden water butt. More complex constructed household rainwater harvesting systems can, however, increase the amount of water 'saved'. They are probably most easily accommodated into new build designs. In the UK, in-house systems typically consist of an underground storage tank, a filter to prevent the entry of leaves and large solids, a smoothing inlet to stop sediment on the bottom of the tank being disturbed, a pump for distributing the harvested water (either to a header tank or directly to appliances) and a suction filter to prevent the uptake of any floating material as the water is drawn up for use. Systems also have an automatic mains top-up device which, depending upon the system, may supply the storage tank, a header tank or the actual appliances with mains water if harvested supplies run short.

3.2 RISK ASSESSMENT

For the purposes of this health impacts examination the following scenario has been considered:

- Each property has a household rainwater system, with an underground tank (such as that shown in Plate 3.1), collecting rainwater from the roof via downpipes, passing through a filter on entry to the tank. The tank receives top up supplies from the mains water when it is running low and the supplies are used for toilet flushing and garden watering. The system operation and its maintenance are the responsibility of the individual householder.

3.2.1 Possible hazards associated with rainwater harvesting

A health hazard is anything that can potentially cause harm (with harm being loss of life, injury, illness and so on). The following hazards have been identified:

- drowning and near-drowning;
- injury; and
- infection.

3.2.2 Exposure assessment

The HIA was conducted on a hypothetical population based on a newly built estate in the South of England, as outlined in Chapter 2 (section 2.5).

Plate 3.1. Typical tank design (Photo: L. Fewtrell)

3.2.2.1 *Drowning and near-drowning*

This will, to some degree, depend upon the type of tanks used. In some countries, Australia for example, above ground rainwater harvesting tanks are common (Cunliffe 1998). In the UK, however, tanks tend to be underground. All such tanks have manhole covers and some prevent actual access to the water by the positioning of the filter (Plate 3.2). Thus, it has been assumed that there can be no exposure and hence no risk of drowning.

3.2.2.2 *Injury*

The scope for injury is determined by user behaviour during maintenance, especially during gutter clearing and tank inspection and cleaning. The former would require ladder access and the latter may present a confined space access risk. There is also the possibility that the manhole cover could be removed during inspection or left unsecured following inspection.

Falls are the most common cause of both fatal and nonfatal unintentional injury (accidents) in the home environment (Dowswell *et al.* 1999; Marshall *et al.* 2005). An important component of this statistic is made up by falls from

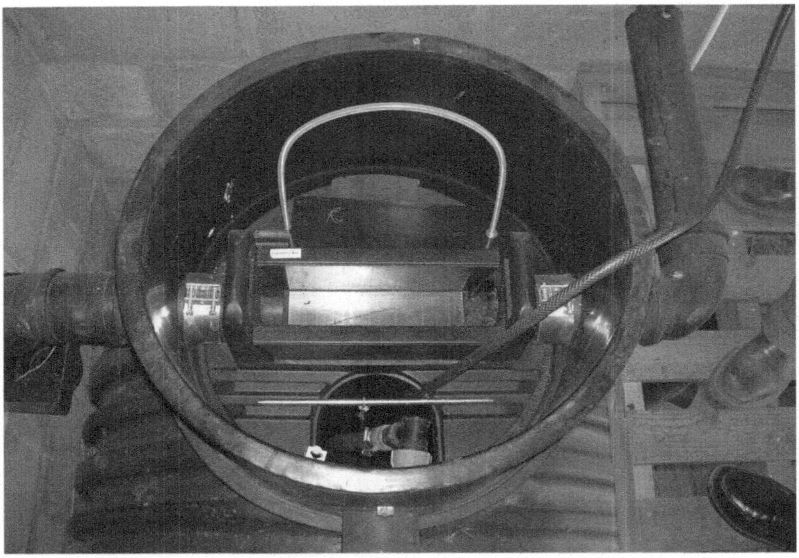

Plate 3.2. Manhole access showing filter (Photo: L. Fewtrell)

ladders (and scaffold), usually as a result of incorrect ladder placement or from excessive reaching (Partridge *et al.* 1998). Examining the risk of injury per hour of exposure to various consumer products, Hayward (1996) found that ladders and scaffolding rated highly, with between 10 and 65 accidents per million hours of use (with scaffold being second in the list, behind electric hedge-trimmers). Ladders and scaffolding were also among the products associated with the longest mean duration of incapacity, with an average incapacity per accident of 18.8 days (Hayward 1996).

In Australia, a study looking at the nature, severity and outcome of injuries sustained from ladder falls found that 83% of those presenting in hospital as a result of a ladder fall were male and that 78% were injured in a non-occupational setting (Tsipouras *et al.* 2001). Over 40% of the accidents were caused by ladder instability. Most of the patients had mild or moderate injuries, but 13% had severe trauma, usually with head, chest or spinal injury. In Denmark, a similar study found that the mean annual rate of ladder falls was 1.18 per 1,000 in males and 0.41/1,000 in females (Faergemann and Larsen 2000). A rise in the annual incidence rates for both men and women was seen with increasing age. About 20% of injuries resulted in hospitalisation for a median of seven days. 50% of the injuries were contusions (bruises) or sprains and about 30% were fractures or dislocations.

In the UK, good surveillance data are available relating to falls including those from ladders. In 1994 and 1995, 73 and 67 people respectively were killed as a result of falling from a ladder or scaffolding in England and Wales. Of these fatalities over 50% occurred at home (as opposed to being occupational injuries) and the percentage of home fatalities was found to increase by age (Dowswell *et al.* 1999). The Home Accident Surveillance System (HASS) maintained by RoSPA is a UK database holding details of home accidents (up to 2002) that were serious enough to warrant a visit to hospital. Data are derived from a sample of admissions to between 16 and 18 hospital accident and emergency departments in the UK, and national estimates are made based on this sample. The data for falls on/from ladders/stepladders for 2000 to 2002 are shown in Table 3.1.

Table 3.1. Incidents involving falls from or onto ladders and stepladders in the home 2000 to 2002 (DTI 2003)

	Number of incidents		
Year	Sample data	National estimate	Multiplier*
2000	1816	32,216	17.64
2001	1903	33,969	17.85
2002	1721	35,281	20.50

* The multiplier was not given in the report, but was calculated from the national estimate and sample data for each year.

Examination of data for 2000 to 2002 revealed 187 incidents involving the use of ladders to access gutters (135 of these were specifically related to cleaning gutters). Using an average of the multipliers (Table 3.1) this suggests that on average 840 people each year will injure themselves cleaning their gutters. As only one of the patients was under the age of 16, the incidence has been calculated based on the adult population (48,047,425 – National Statistics 2005), giving an incidence rate of 0.017/1000 population. From the annual incidence rate it is estimated that 0.062 people per year at the case study site will injure themselves as a result of examining or cleaning their guttering. It has been assumed that householders are aware of their rainwater harvesting system and a proportion of them may clean their gutters more frequently than before as a result of perceived maintenance needs. Thus 20% of incidents have been attributed to rainwater harvesting, resulting in an estimate of 0.012 people sustaining injuries.

Confined space qualitative estimate

A confined space is any space of a substantially closed nature where there is a risk of death or serious injury from hazardous substances or dangerous conditions (e.g. lack of oxygen). Some confined spaces are fairly easy to identify; enclosures with limited openings, such as an underground rainwater harvesting tank (HSE 1997). Data on morbidity and mortality relating to exposure to confined spaces in non-occupational circumstances are not readily available. It is likely that incidents relating to confined spaces (including those arising from occupational exposure) would be coded as W81[1] (confined to or trapped in a low-oxygen environment) or W84 (unspecified threat to breathing). During the three year period (2001 – 2003) there were three deaths coded as W81 in England and Wales and 99 coded as W84 (National Statistics 2001, 2002, 2003), representing annual incidence rates of 0.018/million population and 0.62/million population respectively. Neither of these codes, however, is very specific (e.g., W81 includes 'accidentally shut in refrigerator or other airtight spaces, diving with insufficient air supply' – WHO 2003). The type of tank in the case study is designed to prevent entry (Plate 3.2), so it has been assumed that there is no risk of confined space entry. As with some tank designs, however, entry may be possible a qualitative estimate of (−) has been given to this hazard.

3.2.2.3 Infection

Infection relating to rainwater harvesting could occur via a number of routes, namely:

- Inappropriate ingestion/contact: through ingestion of aerosols produced as a result of toilet flushing; direct ingestion via the garden tap; direct contact through using the garden tap to fill up paddling pools, hot tubs, swimming pools etc., inadvertent ingestion/contact through contamination of drinking-water supplies as a result of cross connections.
- Inhalation of microorganisms within an aerosol: via toilet flushing.
- Microbial contamination of the environment: and subsequent ingestion of garden produce contaminated as a result of watering with rainwater.
- Vector-borne illness: stored water could become a breeding site for mosquitoes.

---

[1] i.e. according to the International Classification of Disease, version 10 (WHO 2003)

The principal pathogen source in harvested rainwater in the UK is likely to be from bird faeces. From an examination of the literature it is clear that a number of different bird species carry a variety of human pathogens, which could be deposited on roofs and washed off into harvested rainwater supplies. The two most commonly studied are *Salmonella* spp. and *Campylobacter* spp. (typically *Campylobacter jejuni*). A recent review (Fewtrell and Kay 2007) suggested that a number of human enteric pathogens have been isolated from rainwater supplies; including *Campylobacter* spp., *Salmonella* spp., *Cryptosporidium* spp. and *Giardia* spp. *Campylobacter* spp. and *Cryptosporidium* spp. were chosen here to quantify the risks of infection. While it is acknowledged that other pathogens may be present in the water the choice of these pathogens is generally conservative (with *Campylobacter* having a lower infectious dose than *Salmonella*, and *Cryptosporidium* having a similar infectious dose to *Giardia*) and the risk from these pathogens was felt to be indicative of infection risk with bacterial and protozoan pathogens.

The following possibilities of infection were examined:

- ingestion of *Campylobacter* spp. from aerosol formed during toilet flushing;
- ingestion of *Campylobacter* spp. and *Cryptosporidium* spp. due to direct ingestion of harvested rainwater or through contamination of drinking water supplies via cross connections; and
- ingestion of garden produce contaminated with *Cryptosporidium* spp.

3.2.2.3.1 Campylobacter spp.

Campylobacter spp. is a bacterial pathogen capable of causing human infection, which is commonly carried by birds such as blackbirds, starlings and gulls (Moore *et al.* 2002; Waldenstrom *et al.* 2002; Broman *et al.* 2002) and has been detected in harvested rainwater supplies (Savill *et al.* 2001; Albrechtsen 2002). *Campylobacter* is the most commonly reported bacterial cause of infectious intestinal disease (gastroenteritis) in England and Wales (HPA 2005). The illness, campylobacteriosis, is characterized by severe diarrhoea and abdominal pain. In some cases chronic sequela (secondary adverse health outcomes), such as Guillain-Barré syndrome, may occur as a result of *Campylobacter* infection (Mead *et al.* 1999).

Dose-response is the quantitative relationship between dose and outcome (e.g. ID_{50} is the number of microbes required to initiate infection in 50% of the exposed population). A dose-response (β-poisson) model for *Campylobacter* spp. has been developed by Medema *et al.* (1996), based on the experimental data reported by Black *et al.* (1988). A duration of 6 days, with a severity weight of 0.086 has been assumed for uncomplicated campylobacteriosis (based on a weight of 0.067 for the majority of cases and 0.39 for the 6% of cases expected to visit their general practitioner). Where there are complications a severity weight of 0.28, with a duration of 365 days has been assumed (adapted from Havelaar *et al.* 2000). A total of 30% of infections are assumed to cause clinical illness and 0.5% of clinical illness is considered to be severe or complicated (based on the rate for campylobacteriosis hospitalisation cited by Mead *et al.* 1999). The case fatality rate for campylobacteriosis is 0.005% (Mead *et al.* 1999), with a median age at death of 78 years (Havelaar *et al.* 2000).

Exposure to Campylobacter spp. through toilet flushing

The data required to estimate the exposure of the case study population to *Campylobacter* through the use of harvested rainwater supplies for toilet flushing are shown in Table 3.2.

It has also been assumed that each day has the same likelihood of *Campylobacter* contamination and that the microorganisms are suspended homogenously in water.

Exposure to Campylobacter spp. through direct ingestion (external tap use)

It has been assumed that during the summer months, and in particular the six week school holiday period, some children may drink water from the garden tap (supplied by harvested rainwater). This assumes that any signs relating to the water not being for drinking are ignored and there is access to the taps. It has also been assumed that children under the age of 5 are either supervised or could not turn on the tap. The data requirements for this analysis are shown in Table 3.3.

Exposure to Campylobacter spp. through cross connections (contaminated potable supplies)

There are no data on the number of cross connections that result from having a dual supply (potable and non-potable) system within the household in the UK. It has, however, been seen in other countries in relation to large scale recycling schemes (e.g. Murray 2005). Thus, it remains a possibility, despite the best efforts of system installation engineers and designers to clearly differentiate between the systems on the basis of pipe colour, dimensions and incompatible fittings. The data requirements for this analysis are shown in Table 3.4.

Table 3.2. Exposure to *Campylobacter* spp. through toilet flushing

Data requirements	Comments
The concentration of *Campylobacter* in the harvested rainwater	Very few studies have attempted to quantify levels of *Campylobacter* spp. in harvested rainwater (presence/absence data only). Savill *et al.* (2001), however, did quantify levels and found a maximum concentration of 0.56 MPN/100ml.
The frequency of contamination of rainwater supplies with *Campylobacter* spp.	Albrechtsen (2002) reported *Campylobacter* spp. in 20% of samples, while Holländer *et al.* (1996) did not detect it in over 140 samples analysed. For this analysis it has been assumed to occur between 0 and 10% of the time (with no allowance made for seasonal variation).
The fraction of these organisms capable of initiating infection	Dose-response model of Medema *et al.* (1996), where α=0.145 and β=7.59
Amount of water swallowed	Flushing a toilet produces an aerosol. The volume of water ejected during a 'typical' flush is unknown but is likely to be small and probably between 1 and 2ml. Only a proportion (say a tenth) of this is likely to reach a susceptible host.
Number of flushes per day	Frequency of toilet use is assumed to be between 3 and 6 times/day, with the range accounting for home workers and those who work away from home (MTP 2006). Children under the age of 3 are not thought to be exposed as they are unlikely to be toilet trained (AAP 2000).
The frequency with which people are exposed to contaminated flush water	It has been assumed that people will be exposed to aerosol 5% of the time (1 flush in 20)

3.2.2.3.2 Cryptosporidium spp.

Cryptosporidium spp. is a protozoan pathogen capable of causing human infection. Cryptosporidiosis commonly produces self-limiting diarrhoea which can sometimes include nausea, vomiting and fever. Although it usually resolves within a week in otherwise healthy people, it can last for a month or more. The severity of illness depends on age and immune status; infections in immunocompromised people can be severe and even life-threatening. The most

Table 3.3. Exposure to *Campylobacter* spp. through direct ingestion (external tap use)

Data requirements	Comments
The level of contamination with *Campylobacter* spp.	See Table 3.2
The frequency of contamination of rainwater supplies with *Campylobacter* spp.	See Table 3.2
The fraction of these organisms capable of initiating infection	See Table 3.2
The amount of un-boiled water consumed	A UK report on water consumption (MEL 1996) reported that the average amount of tap water consumed daily was 1138ml of which un-boiled water accounted for 16.7% (or 190ml)
The number of children likely to drink from external taps	It has been assumed that during the 6 week summer holiday period between 0 and 5% of children aged between 5 and 14 drink from the outside tap on an occasional basis (between 0.5 and 2 times a week) and consume between 50 and 250ml on each occasion

Table 3.4. Exposure to *Campylobacter* spp. through cross connections

Data requirements	Comments
The likely number of households to be affected by cross connections, and the duration of the problem	It has been assumed that between 0 and 0.1% of households are affected each year and that the problem lasts for between 10 and 60 days. Other households are assumed not to be affected.
The level of contamination with *Campylobacter* spp.	0.28 MPN/100ml based on a 50% reduction in the concentration reported by Savill *et al.* 2001 (see Table 3.2) to account for the dilution effect of the drinking water supply
The frequency of contamination of rainwater supplies with *Campylobacter* spp.	See Table 3.2
The fraction of these organisms capable of initiating infection	See Table 3.2
The amount of un-boiled water consumed	See Table 3.3

commonly used dose-response relationship is an exponential model (Teunis *et al.* 1996; WHO 2002) based on data from DuPont *et al.* 1995. A duration of 6 days, with a severity weight of 0.054 has been assumed for the immunocompetent ('normal') population (with 71% of infections progressing to illness), while a duration of 47 days and a severity weighting of 0.13 has been assumed for AIDS patients, where all infection progresses to illness (Havelaar *et al.* 2000). As almost 90% of the case study population classified their health as good, a rate of 0.05% HIV/AIDS prevalence has been assumed, rather than the estimated UK national level of 0.2% (CIA 2005).

Exposure to Cryptosporidium spp. through direct ingestion (external tap use)

Many of the data requirements are the same as those for *Campylobacter* spp. detailed above. The data requirements are outlined in Table 3.5.

Table 3.5. Exposure to *Cryptosporidium* spp. through direct ingestion (external tap use)

Data requirements	Comments
The concentration of *Cryptosporidium* in the harvested rainwater	Albrechtsen (2002) reported *Cryptosporidium* concentrations ranging between 0 and 50/l
The frequency of contamination of rainwater supplies with *Cryptosporidium* spp.	0–10% – as for *Campylobacter*
The fraction of these organisms capable of initiating infection	Dose-response model of Teunis *et al.* (1996) applies, where r = 0.004
The amount of un-boiled water consumed	See Table 3.3
The number of children likely to drink from external taps	See Table 3.3

Exposure to Cryptosporidium spp. through cross connections (contaminated potable supplies)

The requirements to determine the risk from exposure to *Cryptosporidium* spp. as a result of cross connections are shown in Table 3.6.

Table 3.6. Exposure to *Cryptosporidium* spp. through cross connections

Data requirements	Comments
The likely number of households to be affected by cross connections	See Table 3.4
The level of contamination with *Cryptosporidium* spp.	As Table 3.5, but by reduced by 50% to account for dilution by drinking water supply
The frequency of contamination of rainwater supplies with *Cryptosporidium* spp.	See Table 3.5
The fraction of these organisms capable of initiating infection	See Table 3.5
The amount of un-boiled water consumed	See Table 3.4

Exposure to Cryptosporidium spp. through contaminated garden produce

It has been assumed that people in the case study population growing their own fruit and vegetables will water their produce using rainwater supplies, which could on occasion be contaminated with *Cryptosporidium* spp. The products considered to present the greatest risk are salad plants, as the food part of the plant (which is subsequently eaten raw) will be watered directly and, in addition to possible root uptake, may retain any contamination (at least for a period of time). Data requirements for this analysis are outlined in Table 3.7.

No account has been taken of pathogen removal with washing, or of people other than the householders who are growing the produce eating contaminated salad crops (i.e. no assessment of risks from householders who give away some of their produce has been made).

3.2.2.3.3 Other infections and qualitative estimates

Inhalation of microorganisms via aerosols, contact and dermal exposure to rainwater supplies and vector-borne illness were not subject to quantitative microbial risk assessment. *Legionella pneumophila* would be expected to be the principal pathogen of concern in aerosol inhalation, however, this is rarely found in rainwater samples in Europe (e.g. Holländer *et al.* 1996; Albrechtsen 2002; Birks *et al.* 2004) and is unlikely to multiply at the temperatures recorded in underground tanks. Thus this hazard has been given an estimate of (−).

Although there is a possibility that people could be exposed to *Pseudomonas aeruginosa* via rainwater supplies (Holländer *et al.* 1996; Albrechtsen 2002), this is an opportunistic pathogen (Hardalo and Edberg 1997) which only

Table 3.7. Exposure to *Cryptosporidium* spp. through consumption of contaminated garden produce

Data requirements	Comments
The level of contamination with *Cryptosporidium* spp.	See Table 3.5
The frequency of contamination of rainwater supplies with *Cryptosporidium* spp.	See Table 3.5
The fraction of these organisms capable of initiating infection	See Table 3.5
The number of households using harvested rainwater to water food plants eaten raw (e.g. salad crops)	A MORI poll, commissioned by the Royal Horticultural Society estimated that 41% of people grow their own fruit and vegetables (410/1000) and that over a 2-year period 12% of these have attempted to grow lettuce (MORI 2004; RHS 2004). Thus is has been assumed that each year 6% of the fruit and vegetable growing population grow lettuce (i.e. 25/1000). This figure has been applied on a household basis to the study population.
The consumption of such home grown produce	Figures from Northern Ireland (NFS 2005) suggest an average weekly lettuce (and leafy salad) consumption of 25.9 g/person. It has been assumed that families who grow their own vegetables will eat double this amount of their own produce between June and October (i.e. a five month season).
The level of contamination of salad crops	It is assumed that any contamination resulting from watering in the final week of growth will remain on the produce during harvesting

occasionally causes skin problems in healthy people (usually in association with hot tubs or similar facilities – Rasmussen and Graves 1982; CDC 2000) and has, thus, been given a qualitative estimate of (−).

It is unlikely that nutrient levels within an underground rainwater tank would be adequate for the development of mosquito larvae; therefore rainwater harvesting systems are unlikely to play a role in vector-borne disease.

3.2.3 Risk characterization

The estimates relating to injury are determined by multiplying the number of occurrences by the severity weight and duration to calculate the years lived with a disability. Where an occurrence is fatal it has been assumed to occur at the age of eight in children and 45 in adults (unless otherwise noted). The years of life lost are calculated by subtracting the age of death from the UK average (79 based on rounded figures of 77 in men and 81 in women – GAD 2007).

3.2.3.1 Injury

The analysis of exposure (3.2.2.2) estimated that 0.012 people would be injured on an annual basis as a result of cleaning their gutters to maintain their rainwater harvesting system. It has been assumed that no fatalities occur, but that people suffer either from hospitalisation (20%), fractures/dislocations (20%) or bruises/sprains (50%). Table 3.8 shows the severity weights and durations attributed to each of these outcomes (WHO 2005; Stouthard *et al.* 1997).

Table 3.8. Severity weights and duration of incapacitation for ladder-related injuries

Outcome	Severity weight	Duration (days)	Comments
Hospitalisation	0.25	120	Based on double the figures for fractures
Fractures/dislocations	0.125	60	Severity weight based on the average of a break of the radius, ulna, hand bones, ankle, foot bones and dislocation. Duration considered to be three months.
Bruises/sprains	0.064	14	Severity weight based on figures for sprain

Applying these figures to the number of injuries gives a DALY score of 0.0003.

3.2.3.2 Infection

Where possible, parameters for the individual infection exposures were entered into @Risk (Palisade Corporation 2002) as probability distributions rather than point estimates, in order to examine the effects of uncertainty and the assumptions to which the estimate was most sensitive. The data for severity weights and each of the infection exposures are summarised as a series of Tables in the Appendix (Tables A3.1–A3.7).

3.2.3.3 Summary of risk characterization

Table 3.9 summarises the risk characterization for each of the identified hazards applied to the case study population.

Table 3.9. Risk characterization summary (rainwater harvesting)

Hazard	Exposure	Group	Estimate	Annual cases	DALYs
Drowning		All		0	0
Injury	Gutter-related incident				
		Adults		0.012	0.0003
	Confined space incident				
		All	Qualitative		(−)
Illness	Campylobacteriosis from toilet flushing				
		All	Min	3.2×10^{-5}	9.7×10^{-8}
			Mean	0.015	4.6×10^{-5}
			Max	0.11	3.4×10^{-4}
	Campylobacteriosis from direct ingestion (external tap)				
		Children	Min	2.2×10^{-4}	6.8×10^{-8}
			Mean	0.006	1.8×10^{-5}
			Max	0.038	1.1×10^{-4}
	Cryptosporidiosis from direct ingestion (external tap)				
		Children	Min	3.3×10^{-5}	2.9×10^{-8}
			Mean	0.004	3.4×10^{-6}
			Max	0.035	3.1×10^{-5}
	Campylobacteriosis from cross connections				
		All	Min	0	0
			Mean	0.003	9.9×10^{-6}
			Max	0.022	6.5×10^{-5}
	Cryptosporidiosis from cross connections				
		Norm pop	Min	0	0
			Mean	0.002	1.9×10^{-6}
			Max	0.024	2.4×10^{-5}
		Immunocomp	Min	0	0
			Mean	1.4×10^{-6}	2.3×10^{-8}
			Max	1.7×10^{-5}	2.8×10^{-7}
	Respiratory illness via aerosol				
		All	Qualitative		(−)
	Skin infection via dermal contact				
		All	Qualitative		(−)
	Cryptosporidiosis from ingestion of garden produce				
		Norm pop	Min	2.2×10^{-4}	2.2×10^{-7}
			Mean	0.08	8.0×10^{-5}
			Max	0.47	4.6×10^{-4}
		Immunocomp	Min	1.2×10^{-7}	2.0×10^{-9}
			Mean	4.3×10^{-5}	7.3×10^{-7}
			Max	2.5×10^{-4}	4.2×10^{-6}
	Vector-borne illness				
		All		0	0

Norm pop: Normal population
Immunocomp: Immunocompromised population
Min: minimum estimate
Mean: mean estimate
Max: maximum estimate
(−): minor health impact

Figure 3.1 shows a graphical representation of the results of the mean estimate.

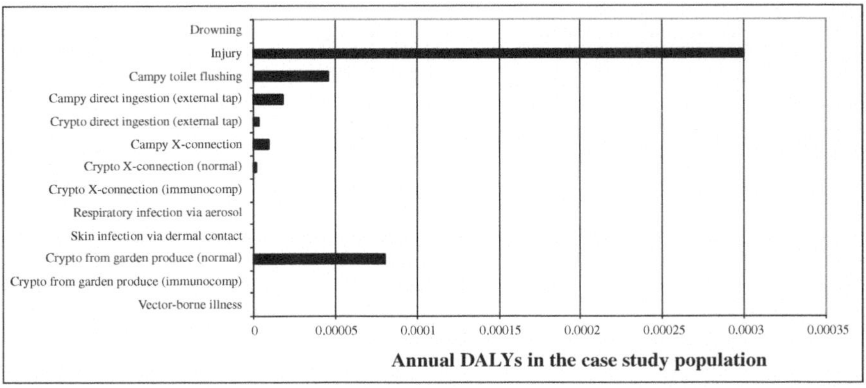

Figure 3.1. Annual DALYs by possible hazard (rainwater harvesting system)

It can be seen from Figure 3.1 that injury has the greatest DALY score, followed by cryptosporidiosis from eating contaminated garden produce and campylobacteriosis from toilet flushing. This information could be used to prioritise risk management strategies, such as an estate rainwater harvesting package where an external company services the system each year, eliminating the need for households to attempt cleaning their gutters. Such a package could actually have in an additional health benefit as, presumably, it would remove the need for any householder to clean their gutters (for what ever reason), which could prevent an additional 0.05 incidents (equivalent to 0.001 DALYs).

3.3 HEALTH IMPACT STATEMENT

This brings together all of the identified health impacts summarised in Table 3.9 (mean estimates) and presents the results on the basis of an overall DALY score.

Total DALY score: 4.59×10^{-4}

Qualitative estimate: 3(−)

3.4 DISCUSSION

The estimated DALY scores from both the mean estimate and the maximum estimate fall well within the screening level and the WHO reference level of risk outlined in Chapter 2. This is clearly shown in Figure 3.2, which compares the annual DALY estimates and the reference levels, on a case study population basis, on a $\log_{10}$ scale.

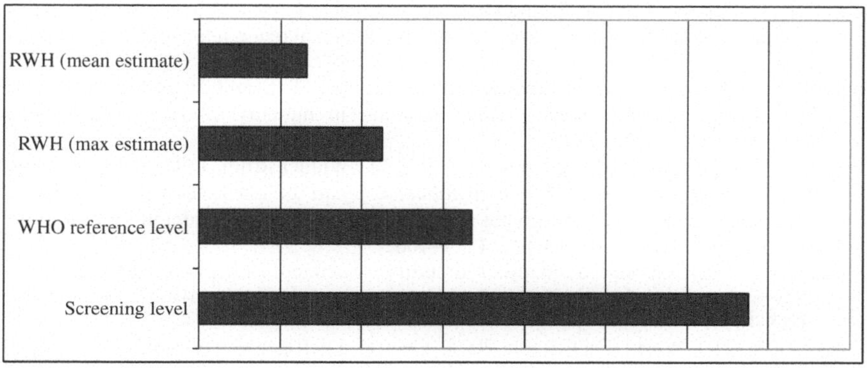

Figure 3.2. Summary of DALY scores ($\log_{10}$ scale)

In order to conduct the HIA, it was necessary to make a number of assumptions, such as the number of people likely to be injured while cleaning their gutters. As far as possible, these were based on data in the literature. Where no data were available (such as the number of children likely to drink from an external tap) documented estimates were made which enable the assumptions to be challenged and (if appropriate) alternatives to be investigated.

For the QMRA, ranges of data were incorporated into the estimates with Monte Carlo analysis used to derive minimum and maximum values. As noted above, even the maximum estimate falls well within both reference levels of risk. It is however, possible to test further some of these assumptions by deliberately choosing elevated levels. For example, ingestion of aerosolized toilet flush water could be increased to 10ml per flush (surely a noticeable and hopefully infeasible amount) and the individual DALY score for that aspect of the assessment still falls within the reference risk levels. Similarly, if the *Campylobacter* spp. concentration of the toilet flush water (derived from the harvested rainwater) is increased to 100/100ml (a level not dissimilar to that seen in polluted river water), the individual DALY score reaches 3.4×10^{-6}, slightly greater than the WHO reference level, but less than the screening level.

3.5 REFERENCES

AAP (2000) Age 3 to 5: Beyond toilet training. American Academy of Pediatrics.

Albrechtsen, H-J. (2002) Microbiological investigations of rainwater and graywater collected for toilet flushing. *Water Science and Technology* **46**, 311–316.

Atherton, F., Newman, C.P.S. and Casemore, D.P. (1995) An outbreak of waterborne cryptosporidiosis associated with a public water supply in the UK. *Epidemiology and Infection* **115**, 123–131.

Birks, R., Colbourne, J., Hills, S. and Hobson, R. (2004) Microbiological water quality in a large in-building, water recycling facility. *Water Science and Technology* **50**(2), 165–172.

Black, R.E., Levine, M.M., Clements, M.L., Hughes, T.P. and Blaser, M.J. (1988) Experimental *Campylobacter jejuni* infection in humans. *Journal of Infectious Diseases* **157**(3), 472–479.

Broman, T., Palmgren, H., Bergstrom, S., Sellin, M., Waldenstrom, J., Danielsson-Tham, M-L. and Olsen, B. (2004) *Campylobacter jejuni* in black-headed gulls (*Larus ridibundus*): prevalence, genotypes and influence on *C. jejuni* epidemiology. *Journal of Clinical Microbiology* **40**(12), 4594–4602.

CDC (2000) Pseudomonas dermatitis/folliculitis associated with pools and hot tubs – Colorado and Maine, 1999–2000. *Morbidity and Mortality Weekly Report* **49**, 1087–1091.

CIA (2005) The World Fact Book 2005. Central Intelligence Agency, USA. http://www.cia.gov/cia/publications/factbook/geos/uk.html accessed 07/12/05

Cunliffe, D. (1998) Guidance on the use of rainwater tanks. National Environmental Health Forum Monographs. Water Series No. 3. Australia.

Dowswell, T., Towner, E., Cryer, C., Jarvis, S., Edwards, P. and Lowe, P. (1999) Accidental falls: fatalities and injuries. An examination of the data sources and review of the literature on preventive strategies. A report prepared for the Department of Trade and Industry URN 99/805.

DTI (2003) 24th (final) report of the home and leisure accident surveillance system. 2000, 2001 and 2002 data. Department of Trade and Industry, London.

DuPont, H.L., Chappell, C.L., Sterling, C.R., Okhuysen, P.C., Rose, J.B. and Jakubowski, W. (1995) The infectivity of *Cryptosporidium parvum* in healthy volunteers. *New England Journal of Medicine* **332**(13), 855–859.

Faergemann, C. and Larsen, L.B. (2000) Non-occupational ladder and scaffold fall injuries. *Accident Analysis and Prevention* **32**(6), 745–750.

Fewtrell, L. and Kay, D. (2007) Microbial quality of rainwater supplies in developed countries. *Urban Water Journal* **4**(4), 253–260.

GAD (2007) Interim life tables 2003–2005. UK. Male and female. Government's Actuary Department, UK Government, London.

Hardalo, C. and Edberg, S.C. (1997) *Pseudomonas aeruginosa*: assessment of risk from drinking water. *Critical Reviews in Microbiology* **23**(1), 47–75.

Havelaar, A.H., de Witt, M.A.S., van Koningsveld, R. and van Kempen, E. (2000) Health burden in the Netherlands due to infection with thermophilic *Campylobacter* spp. *Epidemiology and Infection* **125**, 505–522.

Hayward, G. (1996) Risk of injury per hour of exposure to consumer products. *Accident Analysis and Prevention* **28**, 115–121.

Holländer, R., Bullermann, M., Groß, C., Hartung, H., König, K., Lücke, F-K. and Nolde, E. (1996) [Microbiological and hygienic aspects of the use of rainwater as process water for toilet flushing, garden irrigation and laundering]. *Gesundheitswesen* **58**, 288–293. [Article in German]

HPA (2005) Campylobacter. Health Protection Agency, London. http://www.hpa.org.uk/infections/topics_az/campy/menu.htm Accessed November 2006

HSE (1997) Safe work in confined space. Health and Safety Executive, UK.

Marshall, S.W., Runyan, C.W., Yang, J., Coyne-Beasley, T., Waller, A.E., Johnson, R.M. and Perkis, D. (2005) Prevalence of selected risk and protective factors for falls in the home. *American Journal of Preventive Medicine* **28**(1), 95–101.

Mead, P.S., Slutsker, L., Dietz, V., McCaig, L.F., Bresee, J.S., Shapiro, C., Griffen, P.M. and Tauxe, R.V. (1999) Food-related illness and death in the United States. *Emerging Infectious Diseases* **5**(5), 607–625.

Medema, G.J., Teunis, P.F.M., Havelaar, A.H. and Haas, C.N. (1996) Assessment of the dose-response relationship of *Campylobacter jejuni*. *International Journal of Food Microbiology* **30**, 101–111.

MEL (1996) Tap water consumption in England and Wales: findings from the 1995 national survey. MEL Research Report 9448-01.

Moore, J.E., Gilpin, D., Crothers, E., Canney, A., Kaneko, A. and Matsuda, M. (2002) Occurrence of *Campylobacter* spp. and *Cryptosporidium* spp. in seagulls (*Larus* spp.). *Vector Borne Zoonotic Diseases* **2**(2), 111–114.

MORI (2004) Good Life Factor RHS. Omnibus questions – topline results. MORI/22234. http://www.chelseashow.com/publications/pubs/garden0704/morireportonline.pdf accessed 16/01/06

MTP (2006) BN DW WC: WC design and efficiency – briefing note relating to policy scenario objectives in policy brief. Market Transformation Programme.

Murray, K. (2005) Families forced to drink effluent. *Daily Telegraph, Australia* January 12, 2005.

National Statistics (2001) Mortality statistics – cause. Review of the Registrar General on deaths by cause, sex and age in England and Wales, 2001. Series DH2 no. 28. Office for National Statistics, London.

National Statistics (2002) Mortality statistics – cause. Review of the Registrar General on deaths by cause, sex and age in England and Wales, 2002. Series DH2 no. 29. Office for National Statistics, London.

National Statistics (2003) Mortality statistics – cause. Review of the Registrar General on deaths by cause, sex and age in England and Wales, 2003. Series DH2 no. 30. Office for National Statistics, London.

National Statistics (2005) Population Estimates. http://www.statistics.gov.uk/cci.nugget.asp?id=6

NFS (2005) National Food Survey: Quarterly consumption of household food. Northern Ireland.

Palisade Corporation (2002) Guide to using @Risk: risk analysis and simulation add in for Microsoft Excel. Version 4.5. Palisade Corporation, Newfield, New York, USA. http://www.palisade.com

Partridge, R.A., Virk, A.S. and Antosia, R.E. (1998) Causes and patterns of injury from ladder falls. *Academic Emergency Medicine* **5**(1), 31–34.

Rasmussen, J.E. and Graves, W.H. (1982) *Pseudomonas aeruginosa*, hot tubs, and skin infections. *American Journal of Diseases of Children* **136**, 553–554.

RHS (2004) RHS says dig your way to health in national allotments week. Royal Horticultural Society, UK.

Savill, M.G., Hudson, J.A., Ball, A., Klena, J.D., Scholes, P., Whyte, R.J., McCormick, R.E. and Jankovic, D. (2001) Enumeration of *Campylobacter* in New Zealand recreational and drinking waters. *Journal of Applied Microbiology* **91**, 38–46.

Stouthard, M., Essink-Bot, M., Bonsal, G., Barendregt, J. and Kramers, P. (1997) Disability weights for disease in the Netherlands. Rotterdam: Department of Health, Erasmus University. Cited by VGDHS (1999).

Teunis, P.F.M., van der Heijden, O.G., van der Giessen, J.W.B. and Havelaar, A.H. (1996) The dose-response relation in human volunteers for gastro-intestinal pathogens. RIVM Report 284550002, Bilthoven, Netherlands.

Tsipouras, S., Hendrie, J.M. and Silvapulle, M.J. (2001) Ladders: accidents waiting to happen. *Medical Journal of Australia* **174**(10), 516–519.

VGDHS (1999) Victoria Burden of Disease Study: Morbidity. Victorian Government Department of Human Services Melbourne, Victoria, Australia.

Waldenstrom, J., Broman, T., Carlsson, I., Hasselquist, D., Acherberg, R.P., Wagenaar, J.A. and Olsen, B. (2002) Prevalence of *Campylobacter jejuni, Campylobacter lari*, and *Campylobacter coli* in different ecological guilds and taxa of migrating birds. *Applied and Environmental Microbiology* **68**(12), 5911–5917.

WHO (2002) *Guidelines for Drinking-water Quality. Second Edition. Addendum: Microbiological Agents in Drinking Water*. World Health Organization, Geneva, Switzerland.

WHO (2003) *International Statistical Classification of Diseases and Related Health Problems, 10th Revision*. Version for 2003. World Health Organization, Geneva. (http://www.who.int/classifications/icd/en/ Accessed November 2006)

WHO (2005) Disability weights used in the Global Burden of Disease 1990 study. (http://www3.who.int/whosis/menu.cfm?path=evidence,bod,burden,burden_manual,burden_manual_other&language=english). Accessed November 2006

APPENDIX 3.1 SUMMARY OF INFECTION PARAMETERS

Table A3.1 shows the severity weights, durations and incidence for various forms of infection for both *Campylobacter* spp. and *Cryptosporidium* spp. taken from Havelaar *et al.* (2000a,b) and Mead *et al.* (1999). The remaining Tables outline the @Risk input values and sensitivity analysis data.

Table A3.1. Severity weights, duration and incidence

Infection	Severity	Duration (days)	Incidence*	Age at death
Uncomplicated campylobacteriosis	0.086	6	30%	–
Complicated forms of campylobacteriosis	0.28	365	0.5%	–
Fatal cases of campylobacteriosis	1		0.005%	78
Cryptosporidiosis (normal population)	0.054	6	71%	–
Cryptosporidiosis (immunocompromised)	0.13	47	100%	–
Fatal cases of cryptosporidiosis	1		0.022%	78

* Incidence for uncomplicated campylobacteriosis and cryptosporidiosis in the normal and immunocompromised populations relates to the percentage of people in whom infection goes to clinical illness. The other incidence figures are a proportion of those showing clinical illness.

Campylobacter spp. through toilet flushing

Table A3.2. @Risk input values and sensitivity analysis

Input	Distribution	Mean	Range	Regression value
Volume ingestion (ml)	Normal	0.1	0–0.25	0.59
Campylobacter concentration (/ml)	Log Normal	0.001	0–0.00056	0.46
Frequency of contamination of harvested supplies	Normal	3	0–10	0.42
Number of flushes	Discrete		3–6	0.31

Campylobacter spp. through direct ingestion (external tap use)

Table A3.3. @Risk input values and sensitivity analysis

Input	Distribution	Mean	Range	Regression value
Percentage of children exposed	Normal	2.5	0–5	0.52
Campylobacter concentration (/ml)	Log Normal	0.0025	0–0.0056	0.52
Frequency of contamination of harvested supplies	Normal	3	0–10	0.49
Frequency of drinking from external tap	Normal	7.5	3–12	0.29
Volume ingested (ml)	Normal	125	50–250	0.17

Campylobacter spp. through cross connections (contaminated potable supplies)

Table A3.4. @Risk input values and sensitivity analysis

Input	Distribution	Mean	Range	Regression value
Households affected by cross connections	Discrete		0–2	0.78
Campylobacter concentration (/ml)	Log Normal	0.00125	0–0.0028	0.35
Frequency of contamination of harvested supplies	Normal	3	0–10	0.31
Length of cross connections (days)	Normal	33	10–60	0.03

Cryptosporidium spp. through direct ingestion (external tap use)

Table A3.5. @Risk input values and sensitivity analysis

Input	Distribution	Mean	Range	Regression value
Cryptosporidium concentration (/l)	Log Normal	3	0–50	0.63
Percentage of children exposed	Normal	2.5	0–5	0.45
Frequency of contamination of harvested supplies	Normal	3	0–10	0.42
Frequency of drinking from external tap	Normal	7.5	3–12	0.24
Volume ingested	Normal	125	50–250	0.15

Cryptosporidium spp. through cross connections (contaminated potable supplies)

Table A3.6. @Risk input values and sensitivity analysis

Input	Distribution	Mean	Range	Regression value
Number of households affected by cross connections	Discrete		0–2	0.70
Cryptosporidium concentration (/l)	Log Normal	1.5	0–25	0.43
Frequency of contamination of harvested supplies	Normal	3	0–10	0.29
Length of cross connections (days)	Normal	33	10–60	0.02

Cryptosporidium spp. through contaminated garden produce

Table A3.7. @Risk input values and sensitivity analysis

Input	Distribution	Mean	Range	Regression value
Cryptosporidium concentration (/l)	Log Normal	3	0–50	0.80
Frequency of contamination of harvested supplies	Normal	3	0–10	0.52
Residual water (ml)	Normal	3.75	3–4.5	0.11
Volume of lettuce ingested (g/week)	Normal	51.9	30–90	0.04

REFERENCES

Havelaar, A.H., de Wit, M.A.S., van Koningsveld, R. and van Kempen, E. (2000a) Health burden in the Netherlands due to infection with thermophilic *Campylobacter* spp. *Epidemiology and Infection* **125**, 505–522.

Havelaar, A.H., de Hollander, A.E.M., Teunis, P.F.M., Evers, E.G., van Kranen, H.J., Versteegh, F.M., van Koten, J.E.M. and Slob, W. (2000b) Balancing the risks and benefits of drinking-water disinfection: disability-adjusted life years on the scale. *Environmental Health Perspectives* **108**, 315–321.

Mead, P.S., Slutsker, L., Dietz, V., McCaig, L.F., Bresee, J.S., Shapiro, C., Griffen, P.M. and Tauxe, R.V. (1999) Food-related illness and death in the United States. *Emerging Infectious Disease* **5**(5), 607–625.

4

HIA of sustainable drainage options in the UK

Lorna Fewtrell, David Kay and Richard Kellagher

4.1 INTRODUCTION

Sustainable drainage (SUDS) facilitates the treatment and attenuation of surface water, with the ideal being to mimic, as closely as possible, the natural drainage of the site before development and to treat runoff to remove pollutants (Gilmour and Blackwood 2005). There are a variety of different SUDS techniques (Table 4.1), which are often used in a management train to follow the natural pattern of drainage.

Table 4.1. Structural SUDS (adapted from Revitt *et al.* 2003)

System type	Description
Filter drain	Gravelled area, usually a trench, where stormwater can drain through the gravel to be collected in a pipe
Porous asphalt	Open graded powdered/crushed stone with binder designed to be pervious: high void ratio
Porous paving	Continuous surface with high void content, solid blocks with adjoining infiltration spaces: an associated sub-base reservoir structure provides storage
Sedimentation tank (silt trap)	Symmetrical concrete structure containing appropriate depth of water to assist the settling of suspended solids under quiescent conditions
Filter strip	Grassed or vegetated strip of ground that stormwater flows across
Swale	Vegetated broad shallow channels for transporting stormwater
Soakaway	Underground chamber or rock-filled volume: stormwater soaks into the ground via the base and sides
Infiltration trench	A long thin soakaway
Infiltration basin	Detains stormwater above ground which then soaks away into the ground through the base
Retention pond (balancing pond)	Contain some water at all times and attenuates stormwater runoff
Detention basin	Dry most of the time and able to store rainwater temporarily during wet conditions
Extended detention basin	Dry most of the time and able to store rainwater during wet conditions for up to 24 hours
Lagoon	Pond designed for the settlement of suspended solids
Constructed wetland	Vegetated system with extended retention time

A number of benefits of appropriately designed, constructed and maintained SUDS have been identified (NSWG 2004; Gilmour and Blackwood 2005), which include the following:

- reducing peak flows to watercourses or sewers and potentially reducing the risk of downstream flooding;
- reducing volume of water flowing directly to watercourses or sewers from developed sites;
- improving water quality by removing pollutants from diffuse pollutant sources; and
- improving amenity through the provision of public open space and wildlife habitat.

4.2 RISK ASSESSMENT

The SUDS system assessed consists of a pre-existing wetland, a retention pond, a number of detention basins (which often contain small amounts of water) and six swales (although these are described as linear ponds, as they are more like long, steep-sided ditches – Plate 4.1).

Plate 4.1. Linear pond (Photo: R. Kellagher)

4.2.1 Possible hazards associated with SUDS

The following hazards were identified, in terms of the resident population.

- Drowning and near-drowning: Any pond, wetland or wet detention basin presents a hazard of drowning, near-drowning and submersion, particularly for young children.
- Injury: Falls and slips, allergic reaction/skin problems, from contact with blue green algae or other water-loving toxic plants.

- Infection: Infection from SUDS systems are most likely to result from microbial contamination of the environment (from wildlife and domestic animals attracted to the ponds and wetlands).
- Anxiety: It is possible that anxiety could be caused by a number of things, such as parents worrying about their child drowning and/or concern over an increase in mosquitoes.

4.2.2 Exposure assessment

The HIA was conducted on a hypothetical population based on a newly built estate in the South of England, as outlined in Chapter 2.

4.2.2.1 Drowning and near-drowning

Wetlands, ponds and any area containing standing water (such as a detention basin shortly after a storm) may present a hazard of drowning, especially to children (the group most at risk of drowning). In garden ponds and home-based swimming pools, the group most at risk is young children (especially toddlers). In natural waters away from the home, however, the groups most at risk are older children and young adults, where alcohol is often a factor. There are good data on drowning in the UK, including both routine statistics (e.g. RoSPA 2004) and special reports (e.g. DTI 2000; Kemp and Sibert 1992; Sibert *et al.* 2002). There is, however, no information on drownings in ponds, other than garden ponds, so assumptions relating to SUDS have been based on garden pond data. There are a number of figures given for annual garden pond drownings, including:

- 0.0217/100,000 population for the UK population in 2002 (RoSPA 2004);
- 0.0526/100,000 population for children under the age of 15 in 1988 and 1989 (Kemp and Sibert 1992);
- 0.1004/100,000 population for children under the age of 15 in 1998 and 1999 (Sibert *et al.* 2002); and
- 0.2222/100,000 population for children under the age of five (DTI 2000).

The possible mortality rate due to SUDS ponds is assumed to be represented by 10% of the figure for drowning for the UK whole population (i.e. 0.0021/100,000 population) from RoSPA (2004), applied to the case study population aged between 5 and 15. In addition, severe neurological handicap as a result of near drowning (for the same age group) was assumed to occur in an additional 6% of cases (based on the fatal drowning rate – Kemp and Sibert 1992).

The severity weight of 0.76 is based on 'profound intellectual disability' (VGDHS 1999).

4.2.2.2 Injury

The presence of ponds and wetlands in an area could result in the exposure of a population to plants which cause a toxic or allergic reaction. A number of algal species (blue green algae or cyanobacteria) when present in freshwater are capable of producing toxins. These may mediate their toxic effect through ingestion (i.e. consumption of contaminated water) or through direct water contact. Clearly, direct water contact is the most likely exposure via SUDS. Where dermal contact (excluding ingestion) with cyanobacteria occurs the most likely outcomes are inflammatory skin effects (Fewtrell *et al.* 1994). On rare occasions animal deaths have been reported following ingestion of water containing blue green algal toxins. The most notable incident was in 1989, where the death of 20 sheep and 15 dogs was attributed to ingesting blue green algal scum at Rutland Water (Edney 1990; Lawton and Codd 1991).

The formation of cyanobacterial blooms depends upon the conditions in the pond but environmental conditions which favour bloom formation include moderate to high levels of nutrients (especially phosphorus and nitrate or ammonia), water temperatures between 20–30 °C and a pH between 6 and 9 (Carmichael *et al.* 1985).

Slips and falls may also be related to the presence of ponds and some SUDS structures, especially areas that are muddy or slippery. There are, however, limited data in this area, so no quantitative estimate has been made.

Injury qualitative estimate

There is little information available on the occurrence of blue green algal blooms in existing SUDS ponds in the UK and exposure is likely to be uncommon in the resident population (although this may not be true for maintenance staff). Thus, a qualitative estimate of (−) is suggested to indicate the possibility of a relatively minor and rare health outcome.

There are also limited data with regards to slips and falls relating to SUDS structures. This is an identified hazard, but the risks are considered to be low, therefore a qualitative estimate of (−) is suggested.

4.2.2.3 Infection

The possibility of microbial infection from SUDS is most likely as a result of contamination by animals attracted to the SUDS structures or through contact with insect vectors of disease. A number of possibilities are considered, namely:

- contact with dog faeces;
- contact with rat urine;
- contact with insect vectors; and
- contact with waterfowl faeces.

4.2.2.3.1 Dog faeces

Dog faeces can contain a wide range of faecal pathogens including *Campylobacter* spp., *Salmonella* spp., *Cryptosporidium* spp., *Giardia* spp. and *Toxocara* spp. (Figura 1991; Shukla *et al.* 2006; Bagcigil *et al.* 2007). Although dogs may carry these pathogens and shed them in their faeces often without overt signs of illness, human transmission of most is relatively uncommon (Morse *et al.* 1976; Pedraza-Diaz *et al.* 2001; Xiao *et al.* 2007). Generally, therefore, these pathogens are not a cause for concern, with transmission being rare even in dog owners. However, *Toxocara* spp. may survive in the environment for long periods and be present on the soil surface long after all visible signs of faecal material have disappeared and, thus, are examined in more detail below.

Toxocara spp.

Toxocarosis, sometimes termed toxocariasis, is caused by the pathogen *Toxocara canis* (in dogs) and *T. cati* (in cats), both of which can cause infection in humans, although *T. canis* is thought to be the main species responsible for human infection (Gillespie *et al.* 1991). Toxocara eggs are passed out in the faecal material of infected animals and are very resistant to adverse environmental conditions, often remaining infective for two years or more (Overgaauw 1997a). The eggs are not immediately infective, requiring a period of development in the soil, thus there is no risk of toxocarosis from cleaning up freshly deposited dog faeces ('poop scooping').

Symptoms result from the migration of larval forms around the body and include pneumonitis (inflammation of the air sacs within the lung), chronic abdominal pain, seizures, joint pain/swelling and skin rash (visceral larva migrans). Larvae may also enter the eye and cause loss of vision (ocular larva migrans).

A number of studies have examined the prevalence of *Toxocara* infection in animal populations. In the Netherlands, faecal samples from over 270 dogs and over 230 cats were examined for eggs, levels were quite low (2.9% for dogs, 4.7% for cats), although they were notably higher (21%) in stray cats (Overgaauw 1997b). Far higher levels were reported by O'Lorcain (1994b) in stray dogs and cats found in Dublin, where a prevalence of 82.6% was reported for *T. canis* and 42% for *T. cati*. The highest parasite prevalence rates were found in young animals (O'Lorcain 1994b). In Poland, the prevalence of toxocarosis was examined in dogs, cats and foxes (although it is not clear whether the dogs and cats were pets or stray animals). The highest level (39%) was found in cats. A total of 32% of dogs were infected and 16% of foxes (Luty 2001). In the British Isles, *Toxocara* infection has been estimated to affect 12% of dogs (Rée *et al.* 1984).

A number of authors have examined the level of environmental contamination with *Toxocara* spp., as shown in Table 4.2, although none of these relate to pond areas, anywhere which encourages the presence of dog walkers may have high contamination levels, although (as already noted) infections tend to be significantly lower in owned pets.

Table 4.2. Prevalence of *Toxocara* spp. in environmental samples

Country	Sample Location	Prevalence∗ (s) or (l)	Reference
Czech Republic	Parks	20.4% (l)	Dubna *et al.* 2007
Estonia	Public sand pits	17.8% (s)	Talvik *et al.* 2006
Ireland	Playgrounds	15% (s)	O'Lorcain 1994a
Ireland	Family gardens	38% (l)	Holland *et al.* 1991
Ireland	Parks & open spaces	6% (l)	Holland *et al.* 1991
Italy	Parks	50% (l)	Habluetzel *et al.* 2003
Italy	Playgrounds	63.6% (l)	Giacometti *et al.* 2000
UK	London parks	6.3% (s)	Gillespie *et al.* 1991
UK	London parks	66% (s)	Snow *et al.* 1987
UK	Parks	24.4% (s)	Borg & Woodruff 1973

∗ Prevalence by number of samples (s) or number of locations (l)

According to Rée *et al.* (1984) over 2% of the general population of Britain have been infected with *Toxocara*, although most infections are asymptomatic or unrecognised (Rée *et al.* 1984). Gillespie (2001), estimates asymptomatic toxocarosis to be between 5 to 7% in adults in industrialised countries. There are typically less than ten cases of clinical *Toxocara* infection reported in England and Wales each year (HPA 2007).

As there are so few clinical cases reported in the UK, it is considered to be a negligible risk, which can be further reduced by good hygiene interventions (e.g. poop scooping).

Dog faeces qualitative estimate

The problem of dog faeces is considered to pose a negligible health risk (−), although it may lower the aesthetic value of the area reducing the amenity benefit (see section 4.2.3).

4.2.2.3.2 Rat urine

A number of SUDS structures could, if poorly maintained (especially in terms of allowing a build up of the type of litter which could provide a food source), lead to an increase in rat numbers in areas open to public access. Rats are associated with the disease leptospirosis which is transmitted via contact with infected urine or contaminated water.

Leptospirosis

Leptospirosis is principally a disease of tropical countries, although is does occur in the UK with, typically less than 40 reported cases a year in England and Wales (HPA 2006). In the UK, the most common serogroups are *L. hardjo* (often associated with cattle) and *L. icterohaemorrhagiae* – associated with rats and Weil's disease (Fewtrell *et al.* 1994). Leptospires are shed, often in very large numbers, in the urine of infected animals. Man and animals (man is always an accidental host) become infected either directly through contact with infected urine or indirectly via contaminated freshwater or wet soil. Virulent leptospires can gain entry to the body through cuts and abrasions of the skin and through the mucosal surfaces of the mouth, nose and conjunctiva. Due to the non-specific presentation of leptospirosis, there is almost certainly under-diagnosis of the disease, with mild cases probably being dismissed as flu. However, it is unlikely that a significant number of severe or fatal cases are missed.

Rat urine qualitative estimate

Human cases of leptospirosis are uncommon in the UK and are generally occupational related illnesses such as transmission to farmers from cows, thus, a qualitative estimate of (−) is suggested.

4.2.2.3.3 Insect vectors

Ponds, wetlands and any areas of standing water (even on a relatively temporary basis) may provide a habitat for mosquitoes. Currently, none of the mosquitoes in the UK transmit disease to humans (Medlock *et al.* 2007; Ramsdale and Gunn 2005) some, however, can be biting pests that may trigger an allergic reaction (Ismay *et al.* 2004). In addition, their presence may cause anxiety and annoyance and thus result in a negative impact on mental health and well-being (see section 4.2.2.4).

4.2.2.3.4 Waterfowl faeces

It is likely that ponds and wetlands will attract waterfowl. While this may increase the aesthetic value and perceived amenity of the area, waterfowl may carry a number of pathogens to which people could be exposed (although actual ingestion is likely to be on a very infrequent basis). The following sub-sections outline the likely exposure, while the actual risk of infection is shown in section 4.3.4.2.

A number of studies have shown that waterfowl, such as ducks and Canada geese often carry bacterial pathogens such as *Campylobacter* spp. and *Salmonella* spp. (Faddoul and Fellows 1966; Mitchell and Ridgwell 1971; Luechtefeld *et al.* 1980; Clark 2003; Roser *et al.* 2006). The prevalence of pathogen positive samples is likely to vary according to the pathogen in question, the location, season and type of birds present. For *Campylobacter* spp., prevalences seem to range between zero and 58% (Feare *et al.* 1999; Fallacara *et al.* 2001; Clark 2003). The prevalence of *Salmonella* spp. carriage is generally lower than that for *Campylobacter* spp. (e.g. Fallacara *et al.* 2001), thus *Campylobacter* spp. is likely to present the greater hazard. Exposure will require contact with the pathogen and the possibility of ingestion, thus children (especially those with strong mouthing habits or a tendency to eat soil – geophagy) are most likely to be at risk.

Campylobacter spp.

Campylobacter is the most commonly reported bacterial cause of infectious intestinal disease (gastroenteritis) in England and Wales (HPA 2005). The illness, campylobacteriosis, is characterized by severe diarrhoea and abdominal pain. In some cases chronic sequela (secondary adverse health outcomes), such as Guillain-Barré syndrome, may occur as a result of *Campylobacter* infection (Mead *et al.* 1999).

Dose-response is the quantitative relationship between dose and outcome (e.g. ID_{50} is the number of microbes required to initiate infection in 50% of the exposed population). A dose-response (β-poisson) model for *Campylobacter* spp. has been developed by Medema *et al.* (1996), based on the experimental data reported by Black *et al.* (1988). A duration of 6 days, with a severity weight of 0.086 has been assumed for uncomplicated campylobacteriosis (based on a weight of 0.067 for the majority of cases and 0.39 for the 6% of cases expected to visit their general practitioner), where there are complications a severity weight of 0.28, with a duration of 365 days has been assumed (adapted from Havelaar *et al.* 2000). A total of 30% of infection is assumed to cause clinical illness and 0.5% of clinical illness is considered to be severe or complicated (based on the rate for campylobacteriosis hospitalisation cited by Mead *et al.* 1999). The case fatality rate for campylobacteriosis is 0.005% (Mead *et al.* 1999), with a median age at death of 78 years (Havelaar *et al.* 2000).

The exposure of the population to *Campylobacter* spp. can be estimated from the following:

- approximate number of waterfowl;
- concentration of *Campylobacter* spp. in waterfowl faeces and prevalence of pathogen shedders;
- average daily production of faeces by waterfowl;
- likely number of children playing around the pond; and
- likely amount of waterfowl faeces ingested.

Unfortunately empirical data are not always available and it is often necessary to make a number of assumptions.

Waterfowl numbers

These could be estimated by surveying ponds of a similar nature and assuming that it is appropriate to extrapolate the data to the pond in question. To illustrate the process it has been assumed that the pond supports two swans, approximately 12 ducks and about 20 moorhens.

Waterfowl faeces

There are very limited data on the concentrations of pathogens in waterfowl faeces, although Roser *et al.* (2006) reported concentrations in duck faeces of >0.2 *Salmonella* /g dry weight and 26 *Campylobacter* /g dry weight (mean value 10/g) in Australia. Pathogen shedding is likely to be seasonal, but no data have been found to support this contention and so it has been assumed that a proportion of the waterfowl will shed at this level throughout the year. The prevalence of shedding has been assumed to range between 0 and 58% (see above – 4.2.2.3.4).

Mitchell and Wass (1995) measured the mean daily faecal output in swans at 52 g dry weight, it has been assumed that ducks will produce approximately a fifth of that amount (i.e. approximately 10 g dry weight), and moorhens a tenth. Thus, the supposed waterfowl population of the pond would be expected to produce approximately 324 g dry weight each day. Much of this is likely to be deposited within the pond and it has been assumed that only 20% will be deposited on land. It has been further assumed that there is a low likelihood of humans ingesting this avian faecal material (0.1%), given that it is likely that young children would be supervised and areas of highly contaminated ground are likely to be avoided by most people.

Numbers of children

It is felt that children under the age of two are unlikely to be exposed to waterfowl faeces as they would be expected to be in a pram or pushchair. Above the age of two, children are more likely to be exposed. Older children do not tend to exhibit mouthing behaviour, but are still likely to touch fingers to the mouth on a fairly frequent basis and may be exposed via, for example, a football. It has been assumed that at some stage during the year all children over the age of two will visit the pond and numbers are based on the case study population (adjusted to exclude those less than two years old). The frequency of visits is assumed to range between one and ten (on an annual basis).

Ingestion

A study by Schönning *et al.* (2007) estimated a median soil ingestion level in children of the order of 3 mg per exposure (range 0.01 – 164 mg). This is based on exposure to garden soil, where play is likely to be more extended than around a pond area, thus representing a worst case situation.

4.2.2.4 Anxiety

Anxiety, in its mild form, is essentially a feeling of worry or unease and a decrease in well-being. In relation to SUDS, two possible areas of anxiety have been identified, the fear of a child drowning and concern about an increase in mosquito numbers.

4.2.2.4.1 Drowning and safety

Some parents may have concerns about their child drowning. Safety concerns in relation to SUDS have been investigated by Apostolaki and Jefferies (2005) and were found not to be an issue where ponds were well-established. Although concerns were raised where there were new ponds with limited vegetation. This was also found in a study of a SUDS system at Dunfermline, where barrier planting and the presence of reeds on the shallow margins were found to relieve safety fears (CIRIA 2001). Thus a qualitative estimate of (−) has been assigned.

4.2.2.4.2 Mosquitoes and insect vectors

As mentioned earlier, it is possible that in some areas the number of mosquitoes could increase as a result of using SUDS structures. While most of the mosquito species reported in Great Britain prefer to feed on birds and animals, a number, most notably *Culex pipiens* biotype *molestus*, can be voracious human biters (Snow and Medlock 2006). Where these occur, there is likely to be an increase in annoyance in people living around the area, especially in those who experience allergic reactions to mosquito bites and there may be a fear of mosquito-borne disease, despite the fact that mosquitoes in the UK are not currently believed to transmit human disease (Ramsdale and Gunn 2005; Medlock *et al.* 2007). Although SUDS systems could lead to an increase in mosquito numbers, the voracious human biter (*C. pipiens* biotype *molestus*) prefers underground water (such as flooded basements) so should not increase as a result of SUDS, therefore the main impact is likely to be from a perceived problem. A qualitative estimate of (−) has been assigned.

4.2.3 Health benefits

Well-maintained SUDS structures, especially ponds and constructed wetlands, are likely to have perceived amenity value and a resulting health benefit. The physical environment including the presence of green space (such as that provided by landscaped SUDS structures) has been shown to impact upon mental well-being, with dissatisfaction concerning access to green open spaces being associated with poor mental health (Guite *et al.* 2006). Maas *et al.* (2006) noted that the percentage of green space in people's living environment has a positive association with the perceived general health of residents. Green space may have a double benefit on mental health, both in its own right and also by encouraging physical activity as regular exercise is known to prevent some mental health problems (such as anxiety and depression), ameliorate symptoms (notably anxiety), improve quality of life for people with long term mental health problems and improve mood and levels of subjective well-being (CSIP 2007). Thus, a qualitative estimate of + has been assigned to well-maintained and landscaped SUDS systems

Increased activity can improve physical as well as mental health (SFPHM 1996) and it need not be of vigorous intensity to be beneficial (USDHHS 1996). Physical activity has been shown to reduce the risk of premature mortality in general, and of coronary heart disease, hypertension, colon cancer and diabetes in particular, and is important for the health of muscles, bones and joints (USDHHS 1996), thus, a qualitative estimate of + has been assigned.

It is also recognized that the physical environment is important for community morale and social interaction and it is suggested that the aspects relating to landscaping and encouraging physical activity and social interaction be maximized in any design incorporating SUDS features such as ponds and wetlands.

4.2.4 Risk characterization

4.2.4.1 Drowning and near-drowning

Applying a drowning mortality rate of 0.0021/100,000 population (from section 4.2.2.1) to the case study population aged between 5 and 15, assuming an average age at death of 7 (in the 5 to 9 year age group) and 12 (in the 10 to 14 year age group), results in a DALY score of 8.3×10^{-4}. The DALY score for near-drowning, based on neurological handicap is 3.7×10^{-5}.

4.2.4.2 Infection

Where possible parameters for the exposure estimates were entered into @Risk (Palisade Corporation 2002) as probability distributions rather than point estimates, in order to examine the effects of uncertainty and the assumptions to which the estimate was most sensitive (Table 4.3).

Table 4.3. @Risk input values and sensitivity analysis

Input	Distribution	Mean	Range	Regression value
Frequency of pond visits	Discrete		0–10	0.80
Weight of material ingested (mg)	Normal	3	0.01–164	0.50
Concentration of *Campylobacter* spp. in wildfowl faeces (g)	Log Normal	10	0–26	0.14
Frequency of contamination	Normal	25	0–50	0.06

4.2.4.3 Summary of risk characterization

Table 4.4 summarises the risk characterization for each of the identified hazards applied to the case study population.

Table 4.4. Risk characterization summary (SUDS)

Hazard	Exposure and group	Estimate	Annual cases	DALYs or Qual
Drowning	Children		1.2×10^{-5}	8.3×10^{-4}
Near drowning	Children		7.2×10^{-7}	3.7×10^{-5}
Injury	Contact with blue green algae or other toxic plants			
	All	Qual		(−)
	Slips and fall			
	All	Qual		(−)
Infection	Contact with dog faeces			
	Toxocara (All)	Qual		(−)
	Contact with rat urine			
	Leptospirosis (All)	Qual		(−)
	Contact with wildfowl faeces			
	Campylobacter spp. (Children)	Min	1.4×10^{-6}	4.2×10^{-9}
		Mean	2.3×10^{-4}	6.4×10^{-7}
		Max	8.1×10^{-4}	2.4×10^{-6}
	Contact with insect vectors			
	Mosquito-borne illness (All)		0	0
Mental health	Potential increase in mosquitoes			
	Anxiety (All)	Qual		(−)
	Potential fear of a child drowning			
	Anxiety (Adults)	Qual		(−)
	Potential increase in activity			
	Improved well-being (All)	Qual		+
General health	Potential increase in activity			
	Improved general health (All)	Qual		+

Qual: Qualitative estimate
Min: minimum estimate
Mean: mean estimate
Max: maximum estimate
(−): minor negative health impact
+: health benefit

4.2.5 Risk management

There is a small risk of drowning in water-filled SUDS structures. Good design and best practice can ensure that this is reduced as far as possible, by not locating the ponds immediately adjacent to housing, incorporating shallow sloping sides, encouraging plant growth (barrier planting) to deter entry to the water, ensuring the site is visible to passers-by and providing rescue equipment (if appropriate). There is a possibility that frozen ponds could attract people to venture onto the pond. Where ponds do freeze, it is suggested that signs warning of thin ice are used.

The risk from coming into contact with contaminated waterfowl faeces is very low and the other identified injury and infection hazards (all considered to be minor) can be further controlled by good management practices (e.g. the provision of dog bins) and appropriate action (e.g. in the case of a cyanobacterial bloom – reducing nutrient inputs where feasible, treating the water and erecting warning notices).

Anxiety issues may be more difficult to tackle. However, research has shown that there are few safety concerns relating to well established ponds and mosquito control measures (if necessary) are likely to reduce insect-vector concerns to a minimum.

4.3 HEALTH IMPACT STATEMENT

This brings together all of the identified health impacts and presents the results on the basis of an overall DALY and qualitative estimate.

Overall DALY score (based on mean estimates): 8.67×10^{-4}

Qualitative estimate: 6(−), 2+

4.4 DISCUSSION

This study examined the possible health impact of the use of sustainable drainage on the resident population. The estimate is reliant on a number of assumptions and adaptations/extrapolations of existing data. None of the individual estimates, however, amount to a single annual case within the study population. In addition, the estimate of risk (based on DALY score) was below the reference level of risk suggested by the World Health Organization (WHO 2004) and considerably less than the screening level of risk (5×10^{-5} per person – which equates to a level of 0.273 for the whole case study population), adopted in the current research (see Chapter 2).

The SUDS structures were assumed to be working as designed (not causing localised flooding, for example) and well-maintained. Where this is not the case, the amenity value of the system is likely to be diminished and the overall health impacts may change. Some of the potential physical hazards resulting from poor maintenance are highlighted (such as blue green algal blooms, and infection from contact with dog faeces and rat urine) and currently considered to be of minor/negligible importance, with a qualitative estimate of (−). The impact of poor performance and maintenance could be examined on a theoretical basis, although it could be argued that a badly maintained and aesthetically poor site is unlikely to be seen as an amenity and, thus, public exposure would decrease. Anxiety and annoyance, on the other hand, would probably increase.

With the use of certain SUDS structures, there are several identified health benefits such as increased well-being and the opportunity for increased physical activity. Ideally, these should be maximised, especially in areas where other green space is lacking.

In this assessment, no consideration has been given to staff maintaining the SUDS system. For the most part, the nature of the associated occupational hazards are likely to be similar (with the possibility of additional injury hazards; such as those deriving from strimmer use, physical labour and so on), although the exposure is clearly likely to be different. It would be expected, for example, that maintenance staff with have more extensive and closer contact with the plants and water leading to an increased risk from insects and contaminated water. Even where insects are not acting as disease vectors encountering stinging or biting insects in any number is likely to be a nuisance at best and, where an allergic reaction occurs, potentially more serious. Most risks can be mitigated by providing training (including sensible hygiene precautions) and personal protective equipment.

The assessment has been presented as an overall risk to the resident population. From a development and implementation view-point it would be useful to be able to assess the individual SUDS components (for which available data are currently lacking). This would allow comparisons between, for example, a periodically wet basin (where the attraction to the water might be greater than in a permanently wet feature), a wet pond and a ditch. Additional research could investigate the factors that influence risk (such as the proximity of the structure to houses and the steepness of the banks) to enable the most appropriate design and configuration of structures (from a health impact perspective) and effective mitigation strategies to be employed.

4.5 REFERENCES

Apostolaki, S. and Jefferies, C. (2005) Social impacts of stormwater management techniques including river management and SUDS. Final report. Reference SUDS01 SNIFFER.

Bagcigil, A.F., Ikiz, S., Dokuzeylu, B., Basaran, B., Or, E. and Ozgur, N.Y. (2007) Fecal shedding of *Salmonella* spp. in dogs. *Journal of Veterinary Medical Science* **69**(7): 775–777.

Black, R.E., Levine, M.M., Clements, M.L., Hughes, T.P. and Blaser, M.J. (1988) Experimental *Campylobacter jejuni* infection in humans. *Journal of Infectious Diseases* **157**(3), 472–479.

Borg, O.A. and Woodruff, A.W. (1973) Prevalence of infective ova of *Toxocara* species in public places. *British Medical Journal* **4**, 470–472.

Carmichael, W.W., Jones, C.L.A., Mahmood, N.A. and Theiss, W.C. (1985) Algal toxins and water-based diseases. *CRC Critical Reviews in Environmental Control* **15**, 275–313.

CIRIA (2001) Dunfermline Eastern Expansion project, SUDS case study, 18th June 2001. (http://www.ciria.org.uk/SUDS/cs_dunferline_eastern_expansion.htm) accessed 10/05/05

Clark, L. (2003) A review of pathogens of agricultural and human health interest found in Canada geese. In *Proceedings of the 10th Wildlife Damage Management Conference*. (eds K.A. Fagerstone and G.W. Witmer), pp. 326–334.

CSIP (2007) Mental well-being impact assessment: a toolkit. Care Services Improvement Partnership, North West Development Centre.

DTI (2000) Drowning accidents in the garden involving children under five. Department of Trade and Industry: London.

Dubna, S., Langrova, I., Jankovska, I., Vadlejch, J., Pekar, S., Napravnik, J. and Fechtner, J. (2007) Contamination of soil with *Toxocara* eggs in urban (Prague) and rural areas in the Czech Republic. *Veterinary Parasitology* **144**(1–2), 81–86.

Edney, A.T.B. (1990) Toxic algae. *Veterinary Record* 27.10.90, 434–435.

Faddoul, G.P. and Fellows, G.W. (1966) A five-year survey of the incidence of salmonellae in avian species. *Avian Diseases* **10**, 296–304.

Fallacara, D.M., Monahan, C.M., Morishita, T.Y. and Wack, R.F. (2001) Fecal shedding and antimicrobial susceptibility of selected bacterial pathogens and a survey of intestinal parasites in free-living waterfowl. *Avian Diseases* **45**(1), 128–135.

Feare, C.J., Sanders, M.F., Blasco, R. and Bishop, J.D. (1999) Canada goose (*Branta canadensis*) droppings as a potential source of pathogenic bacteria. *Journal of the Royal Society for the Promotion of Health* **119**(3), 146–155.

Fewtrell, L., Godfree, A., Jones, F., Kay, D. and Merrett, H. (1994) Pathogenic microorganisms in temperate environmental waters. Samara Publishing Limited, Tresaith.

Figura, N. (1991) *Campylobacter* spp. isolated from dog faeces. *Lancet* **338**, 1403.

Giacometti, A., Cirioni, O., Fortuna, M., Osimani, P., Antonicelli, L., Del Prete, M.S., Riva, A., D'Errico, M.M., Petrelli, E. and Scalise, G. (2000) Environmental and serological evidence for the presence of toxocariasis in the urban area of Ancona, Italy. *European Journal of Epidemiology* **16**(11), 1023–1026.

Gillespie, S. (2001) *Toxocara*: dogwalking and playing fields. *British Journal of Sports Medicine* **35**, 6–7.

Gillespie, S.H., Pereira, M. and Ramsay, A. (1991) The prevalence of *Toxocara canis* in soil samples from parks and gardens in the London area. *Public Health* **105**, 335–339.

Gilmour, D. and Blackwood, D. (2005) *Review of the SUDS concept, use and legislative framework in the UK*. WaND Briefing Note 27.

Guite, H., Clark, C. and Ackrill, G. (2006) The impact of the physical and urban environment on mental well-being. *Public Health* **120**, 1117–1126.

Habluetzel, A., Traldi, G., Ruggieri, S., Attili, A.R., Scuppa, P., Marchetti, R., Menghini, G. and Esposito, F. (2003) An estimation of *Toxocara canis* prevalence in dogs, environmental egg contamination and risk of human infection in the Marche region of Italy. *Veterinary Parasitology* **113**(3–4), 243–252.

Havelaar, A.H., de Witt, M.A.S, van Koningsveld, R. and van Kempen, E. (2000) Health burden in the Netherlands due to infection with thermophilic *Campylobacter* spp. *Epidemiology and Infection* **125**, 505–522.

Holland. C., O'Connor, P., Taylor, M.R., Hughes, G., Girdwood, R.W. and Smith, H. (1991) Families, parks, gardens and toxocariasis. *Scandinavian Journal of Infectious Diseases* **23**(2), 225–231.

HPA (2005) Campylobacter. Health Protection Agency, London. http://www.hpa.org.uk/infections/topics_az/campy/menu.htm (assessed November 06)

HPA (2006) Leptospirosis – general information. Health Protection Agency, London. http://www.hpa.org.uk/infections/topics_az/zoonoses/leptospirosis/gen_info.htm (accessed 06/12/06).

HPA (2007) Toxocara – background information. Health Protection Agency, London. http://www.hpa.org.uk/infections/topics_az/zoonoses/toxocarosis/geninfo.htm#data (accessed 16/02/07).

Ismay, J.W., Mann, D.J. and Schulten, B. (2004) Report on the survey for mosquito winter breeding sites in the area surrounding Mogden STW in February 2004. Report commissioned by Thames Water Utilities Ltd.

Kemp, A. and Sibert, J.R. (1992) Drowning and near drowning in children in the United Kingdom: lessons for prevention. *British Medical Journal* **304**, 1143–1146.

Lawton, L.A. and Codd, G.A. (1991) Cyanobacterial (blue-green algal) toxins and their significance in UK and European Waters. *Journal of the Institution of Water and Environmental Management* **5**(4), 460–465.

Luechtefeld, N.A.W., Blaser, M.J., Reller, L.B. and Wang, W-L.L. (1980) Isolation of *Campylobacter fetus* subsp. *jejuni* from migratory waterfowl. *Journal of Clinical Microbiology* **12**(3), 406–408.

Luty, T. (2001) Prevalence of species of *Toxocara* in dogs, cats and red foxes from the Poznan region, Poland. *Journal of Helminthology* **75**(2), 153–156.

Maas, J., Verheij, R.A., Groenewegen, P.P., de Vries, S. and Spreeuwenberg, P. (2006) Green space, urbanity and health: how strong is the relation? *Journal of Epidemiology and Community Health* **60**, 587–592.

Mead, P.S., Slutsker, L., Dietz, V., McCaig, L.F., Bresee, J.S., Griffen, P.M. and Tauxe, R.V. (1999) Food-related illness and death in the United States. *Emerging Infectious Diseases* **5**(5), 607–625.

Medema, G.J., Teunis, P.F.M., Havelaar, A.H. and Haas, C.N. (1996) Assessment of the dose-response relationship of *Campylobacter jejuni*. *International Journal of Food Microbiology* **30**, 101–111.

Medlock, J.M., Snow, K.R. and Leach, S. (2007) Possible ecology and epidemiology of medically important mosquito-borne arboviruses in Great Britain. *Epidemiology and Infection* **135**, 466–482.

Mitchell, S.F. and Wass, R.T. (1995) Food consumption and faecal deposition of plant nutrients by black swans (*Cygnus atratus Latham*) in a shallow New Zealand lake. *Hydrobiologia* **306**(3), 189–197.

Mitchell, T.R. and Ridgwell, T. (1971) The frequency of salmonellae in wild ducks. *Journal of Medical Microbiology* **4**, 359–361.

Morse, E.V., Duncan, M.A., Estep, D.A., Riggs, W.A. and Blackburn, B.O. (1976) Canine salmonellosis: a review and report of dog to child transmission of *Salmonella enteritidis*. *American Journal of Public Health* **66**(1), 82–83.

NSWG (2004) Interim code of practice for sustainable drainage systems. National SUDS Working Group. July 2004 ISBN 0-86017-904-4 – cited by Gilmour & Blackwood, 2005

O'Lorcain, P. (1994a) Prevalence of *Toxocara canis* ova in public playgrounds in the Dublin area of Ireland. *Journal of Helminthology* **68**(3), 237–241.

O'Lorcain, P. (1994b) Epidemiology of *Toxocara* spp. in stray dogs in Dublin, Ireland. *Journal of Helminthology* **68**(4), 331–336.

Overgaauw, P.A. (1997a) Aspects of *Toxocara* epidemiology: toxocarosis in dogs and cats. *Critical Reviews in Microbiology* **23**(3), 233–251.

Overgaauw, P.A. (1997b) Prevalence of intestinal nematodes of dogs and cats in The Netherlands. *Veterinary Quarterly* **19**(1), 14–17.

Palisade Corporation (2002) Guide to using @Risk: risk analysis and simulation add in for Microsoft Excel. Version 4.5. Palisade Corporation, Newfield, New York. http://www.palisade.com accessed April 14 2008.

Pedraza-Diaz, S., Amar, C., Iversen, A.M., Stanley, P.J. and McLaughlin, J. (2001) Unusual *Cryptosporidium* species recovered from human faeces: first description of *Cryptosporidium felis* and *Cryptosporidium* 'dog type' from patients in England. *Journal of Medical Microbiology* **50**, 293–296.

Ramsdale, C.D. and Gunn, N. (2005) History of and prospects for mosquito-borne disease in Britain. *European Mosquito Bulletin* **20**, 15–31.

Rée, G.H., Voller, A. and Rowland, H.A.K. (1984) Toxocariasis in the British Isles 1982-3. *British Medical Journal* **288**, 628–629.

Revitt, M., Ellis, J.B. and Scholes, L. (2003) Literature review – BMPs in Europe. Daywater Project under EU RTD 5th Framework Programme Contract No EVK1-CT-2002-00111. Adaptive Decision Support System (ADSS) for the Integration of Stormwater Source Control into Sustainable Urban Water Management Strategies WP5/T5.1/D5.1 – PU - cited by Gilmour & Blackwood, 2005.

Roser, D.J., Davies, C.M., Ashbolt, N.J. and Morison, P. (2006) Microbial exposure assessment of an urban recreational lake: a case study of the application of new risk-based guidelines. *Water Science and Technology* **54**(3), 245–252.

RoSPA (2004) UK drowning statistics – 2002. The Royal Society for the Prevention of Accidents: Birmingham.

Schönning, C., Westrell, T., Stenström, T.A., Arnbjerg-Nielsen, K., Hasling, A.B., Høibye, L. and Carlsen, A. (2007) Microbial risk assessment of local handling and use of human faeces. *Journal of Water and Health* **5**(1), 117–128.

SFPHM (1996) Health related physical activity. Scottish Forum for Public Health Medicine.

Shukla, R., Giraldo, P., Kraliz, A., Finnigan, M. and Sanchez, A.L. (2006) *Cryptosporidium* spp. and other zoonotic enteric parasites in a sample of domestic dogs and cats in the Niagara region of Ontario. *Canadian Veterinary Journal* **47**, 1179–1184.

Sibert JR, Lyons RA, Smith BA, Cornall P, Sumner V, Craven MA, Kemp AM (2002) Preventing deaths by drowning in the United Kingdom: have we made progress in 10 years? Population based incidence study. *British Medical Journal* **324**, 1070–1071.

Snow, K.R., Ball, S.J. and Bewick, J.A. (1987) Prevalence of *Toxocara* species in the soil of five east London parks. *Veterinary Record* **120**, 66–67

Snow, K. and Medlock, J. (2006) The potential impact of climate change on the distribution and prevalence of mosquitoes in Britain. *European Mosquito Bulletin* **21**, 1–10.

Talvik, H., Moks, E., Magi, E., Jarvis, T. and Miller, I. (2006) Distribution of *Toxocara* infection in the environment and in definitive and paratenic hosts in Estonia. *Acta Veterinaria Hungarica* **54**(3), 399–406.

USDHHS (1996) Physical activity and health. A report of the Surgeon General. United States Department of Health and Human Services, USA.

VGDHS (1999) Victoria Burden of Disease Study: Morbidity. Victorian Government Department of Human Services, Melbourne, Victoria, Australia.

WHO (2004) Guidelines for Drinking-water Quality. Third Edition. Volume 1. Recommendations. World Health Organization, Geneva.

Xiao, L., Cama, V.A., Cabrera, L., Ortega, Y., Pearson, J. and Gilman, R.H. (2007) Possible transmission of *Cryptosporidium canis* among children and a dog in a household. *Journal of Clinical Microbiology* **45**(6), 2014–2016.

5

HIA of greywater reuse in the UK

Lorna Fewtrell, David Kay, Paul Jeffrey and Bruce Jefferson

5.1 INTRODUCTION

Greywater from domestic properties is usually defined as water originating from the mains potable supply following use for bathing, washing, washing dishes or laundering clothes (WRAS 1999). Although, technically, greywater is any water from the household that is not blackwater (i.e. derived from toilets or bidets), it is usual to exclude waste from dish washers and kitchen sinks (certainly from simple in-house systems) because these sources will contain fats, grease and oils which can cause fouling and result in blockages and odour problems.

In the UK, greywater is usually utilised for non-potable uses such as toilet flushing and garden watering and many of the issues relating to greywater use are similar to those for rainwater harvesting (see Chapter 3), although greywater is susceptible to degradation during storage because of its organic loading and it can be stored for less time than harvested rainwater. Dixon *et al.* (1999a)

© 2008 IWA Publishing. *Health Impact Assessment for Sustainable Water Management.* Edited by Lorna Fewtrell and David Kay. ISBN: 9781843391333. Published by IWA Publishing, London, UK.

examined changes in stored untreated greywater and found that while storing for 24 hours may improve the water quality (through settlement of organic particles), storage beyond 48 hours was likely to lead to depleted dissolved oxygen levels and potential aesthetic problems.

Greywater use is well established in a number of countries, especially Japan, USA and Australia (Dixon *et al.* 1999b), although not always for domestic purposes. Studies monitoring the performance and acceptability of single-house greywater systems in the UK, seem to suggest that while user perception is quite positive there are issues relating to the reliability of the systems which may undermine the acceptability, negate water savings and potentially lead to negative health impacts (EA 2000; Leggett *et al.* 2001; Hills *et al.* 2003).

5.1.1 Greywater quality

5.1.1.1 Quality prior to treatment

The quality of greywater, prior to treatment, depends principally on the source of the water and the household composition. The main contaminants in greywater include:

- human faecal matter;
- animal faecal matter (from pets);
- hair and skin;
- fats and grease from washing;
- dirt and grit;
- detergents and soap;
- cleaning products (such as bleach); and
- other materials or chemicals discarded in the sink or bath (Leggett *et al.* 2001).

Rose *et al.* (1991) examined the microbial quality of greywater from various sources. They found that total and faecal coliform levels were considerably higher in shower/bath water (10^5 and 6×10^3 cfu/100ml, respectively) than in laundry wash (199 and 55 cfu/100ml) or rinse water (126 and 25 cfu/100ml). Friedler (2004) also reported the highest faecal coliform concentrations from bath and shower water (4×10^6 cfu/100ml) and he found similar levels in washing machine effluent, which is in contrast to the findings of Rose *et al.* (1991) outlined above. The lowest average (believed to be an arithmetic mean) faecal coliform concentration was found in greywater derived from wash basins.

Rose *et al.* (1991) found that in greywater from all sources, coliform levels were low in households without children (typically < 100 cfu/100ml). In households with young children, however, levels were significantly higher, with faecal coliform levels averaging 3.2×10^5 cfu/100ml. Faecal coliform levels were also shown to fluctuate widely within a single household over a ten week period, ranging from 1.82×10^4 to 7.94×10^6 cfu/100ml. In contrast, total coliforms and total bacterial populations did not vary greatly and were quite stable.

Ottoson and Stenström (2003a) claim that focusing on bacterial indicator numbers in greywater will lead to an over-estimation of faecal load within greywater as the indicators may be derived from a number of sources. To try to counter this problem for an estimation of risk, they also measured coprostanol and cholesterol concentrations in greywater as well as the more usual indicator organisms (total coliforms, *E. coli* and intestinal enterococci). Coprostanol is a chemical marker of faecal contamination and is produced as a result of the action of intestinal microflora on cholesterol. The measured indicator concentrations gave widely different estimates of the daily faecal contamination of greywater (i.e. faecal load in g/person/day), ranging from 65 g/person/day based on *E. coli* measurements to 0.04 g/person/day based on coprostanol concentrations. They concluded that the best measure was probably provided by intestinal enterococci (5.4 g/person/day), noting that coprostanol is likely to underestimate faecal load where young children contribute to the loading (as not all children excrete coprostanol).

Birks *et al.* (2004) examined greywater quality from hand basins at the Millennium Dome in London, prior to treatment. Median concentrations (cfu/100ml) were as follows; total coliforms -6.5×10^5, *E. coli* -7×10^3 (max concentration $>2 \times 10^6$), intestinal enterococci – $>2 \times 10^2$. Three samples were taken for pathogen analysis at a time of high visitor numbers (maximising the likelihood of an infected person using the facilities). *E. coli* O157, *Campylobacter* spp., *Salmonella* spp., *Legionella pneumophila* – serogroup-1 (the strain most usually associated with infections) and *Shigella* spp. were not detected in the samples. Samples were, however, positive for *Cryptosporidium*, *Giardia* and *Legionella pneumophila* serogroups 2–14. In the USA, Burrows *et al.* (1991) also examined greywater for a number of pathogens. The greywater was derived from shower water at field military facilities and was analysed for *Candida albicans*, *Ps. aeruginosa*, and *Staphylococcus aureus* (chosen as they commonly occur in the mouth, nose and throat). Neither *C. albicans*, nor *Ps. aeruginosa* were detected, although *S. aureus* was detected in the range 1–5×10^5 cfu/ml.

Palmquist and Hanæus (2005) measured a number of hazardous chemical substances in raw greywater (derived from all sources except toilets;

i.e. including kitchen wastewater) and blackwater samples from ordinary Swedish households. They identified 68 of their selected 105 hazardous substances in greywater. The concentrations detected, however, were very low. A number of the metals detected are listed in the WHO drinking-water quality guidelines (WHO 2004), however, in each case, the concentrations in greywater were less than drinking-water quality guidelines (e.g. cadmium concentration in greywater: 0.1μg/l; drinking-water guideline value: 3μg/l).

5.1.1.2 Potential for the survival of pathogens in untreated greywater

Rose *et al.* (1991) seeded greywater with *Salmonella typhimurium* and *Shigella dysenteriae* and found that they could persist for several days. Ottoson and Stenström (2003b) studied the survival of pathogenic bacteria (*Salmonella* and *Campylobacter*) seeded into sediments from a greywater treatment system and found that the bacteria died off rapidly under normal conditions, suggesting that the normally occurring microbiota suppress pathogen growth. It should be noted, however, that these bacteria are sensitive to disinfection and would not be expected to survive chemical treatment.

Noah (2002) has suggested that where a communicable disease, such as hepatitis, flu or measles, has been diagnosed within the family, greywater use should be discontinued and the greywater diverted into the sewerage system, until the affected person is no longer infectious, to reduce the possibility of disease transmission.

5.1.1.3 Treatment options

The type of treatment the greywater is subjected to prior to reuse can vary and depends upon the scale of the installation, i.e. single versus multiple houses or apartments. Typically, single domestic systems consist of coarse filtration (to remove hair etc. via a metal strainer) and simple chemical disinfection. Larger systems are likely to be more elaborate and may include a number of treatment stages including physical, chemical and biological treatment (EA 2000).

Options for treatment include the following (some of which may be used in combination).

- Disinfection: there are a number of disinfectants available that can be used for greywater treatment. In-house systems generally use doses of chlorine or bromine often dispensed in slow release blocks or by liquid dosing. These are designed to produce a low greywater retention time to avoid degradation and odour problems.

- Physical and biological systems: such as rapid- or slow-sand filters, membrane bioreactors (MBR), biological aerated filers, constructed wetlands, reedbeds and green roofs.
- Nutrient addition as part of biological treatment.

5.1.1.4 Microbial quality post treatment

In a monitoring trial conducted by the Environment Agency (EA) on ten houses retrofitted with greywater systems, most samples taken at the point of use (i.e. the toilet cistern) contained few or zero faecal coliforms (EA 2000), although occasional high total coliform levels were recorded. Where the disinfection system had been deliberately turned off or where it malfunctioned, however, faecal coliform levels ranged between 15,000/100ml (where disinfection was through manual dosing of household cleaning products) to 35,000/100ml (where the disinfection was inadequate, possibly due to low disinfectant stores). In the home where the disinfectant system had been deliberately switched off (due to excessively high levels of chlorine in the header tank, with the result that the loft space became unpleasant to enter), it is believed that dogs washed in the bath may have contributed to the high faecal coliform level.

Hills *et al.* (2003) did not detect faecal coliforms or *E. coli* in samples taken from the toilet cistern. A small amount of pathogen testing was conducted during the study; two samples from one of the houses were found to contain low levels of *Aeromonas* during a period when the disinfection system had failed. None of the samples contained *Salmonella* or *Legionella* spp.

Nolde (1999) studied two different greywater systems in Germany (one using a rotary biological contactor and UV disinfection, the other a fluidized-bed reactor and UV disinfection) and found bacterial concentrations in treated samples were generally low, often below the limit of detection, although *Ps. aeruginosa* was occasionally detected. Nolde (1999) also challenged one of the systems with faeces from baby nappies and found that the system performed well and microbial concentrations were satisfactory (i.e. total coliforms $<$ 100/ml; faecal coliforms $<$ 10/ml and *Ps. aeruginosa* $<$ 1/ml).

In Australia, Grant *et al.* (2006) reported monitoring results for greywater from 100 apartments subjected to a MBR followed by UV disinfection. *E. coli* was detected (maximum concentration 1 mpn/100ml) on two occasions post MBR, although this is attributed to experimental error as subsequent testing did not confirm the initial results. Total coliforms were also detected post MBR but this was thought to be due to regrowth rather than membrane penetration.

5.2 RISK ASSESSMENT

For the purposes of the HIA a household level vertical flow constructed wetland (reedbed), followed by chlorine disinfection was considered. The hypothetical system examined takes water from showers, baths and wash basins (and excludes water from the kitchen sink, dishwasher and washing machine). Although the reused water could be used for toilet flushing, the system was assumed to supply only water for external purposes, such as garden watering, and car washing (the hazard presented by cross connections is explored in Chapter 3 on a rainwater harvesting system). There is a maintenance agreement in place, where systems are tested on a six monthly basis and appropriate maintenance conducted. Householders are responsible for checking and cleaning the coarse filters and for checking the disinfection system (although supplies are designed to last throughout the maintenance period).

A vertical flow constructed wetland (VFCW) is essentially a reedbed and, thus, a biological system, which will contain fungi, algae, bryophytes (mosses and liverworts) and aquatic invertebrates as well as reeds. The reeds provide oxygen and surfaces on which other organisms can grow and proliferate. Greywater is applied in a batch, flooding the bed and then draining through the system and out to a storage tank. The bed includes perforated pipework which allows oxygen to penetrate the bed.

It has been assumed that individual household level the greywater is coarse screened (to remove hairs etc.) and then stored in an underground tank (with a pipe allowing overflow to the sewer). The water is pumped to the reedbed. The biologically treated water is stored in a further tank (with a sewer overflow pipe) and subject to chlorine disinfection, before being available for unrestricted external use, via a tap marked 'not for drinking'.

5.2.1 Possible hazards associated with greywater reuse

The following possible hazards were identified.

- Drowning and near-drowning: Tanks and wetlands may present a hazard of drowning and near-drowning, particularly for young children.
- Injury: This could be maintenance-related, for example electrocution from a submerged pump.
- Chemical intoxication: As a result of exposure to the chlorine disinfectant.
- Anxiety: It is possible that where the system is not running properly and is resulting in noxious smells and/or deposits on the soil, some people may find this distressing and develop anxiety and health concerns.

- Infection: Infection as a result of greywater reuse could, in theory, occur via a number of routes: inappropriate ingestion/contact; microbial contamination of the environment; and vector-borne illness.

5.2.2 Exposure assessment

The HIA was conducted on a hypothetical population based on a newly built estate in the South of England, as outlined in Chapter 2. This section assesses the possible exposures to the various hazards outlined above.

5.2.2.1 Drowning and near-drowning

Constructed wetlands may present a hazard of drowning, especially in young children similar to that presented by garden ponds, with toddlers being the greatest 'at risk' group (Brenner 2002). Kemp and Sibert (1992) examined drowning and near-drowning incidents in children, under the age of 15, in the UK for 1988 and 1989. Over 7% of the deaths and 30% of the near-drowning occurred in garden ponds. Of the near-drowning incidents 6% resulted in a severe neurological handicap. Ten years on, drowning figures were re-examined (Sibert *et al.* 2002) and drowning in garden ponds had increased from 11 in 1988-9 to 21 in 1998-9, a significant increase. Interestingly, overall there was a significant decrease in the numbers of drownings, with decreases seen at all sites with the exception of garden ponds.

The Department of Trade and Industry (DTI 2000) examined drowning accidents in the garden in children under the age of five. Between 1992 and 1999, there were 90 fatal garden drownings in the UK, of which nearly 70% (62) involved garden ponds. The average number of cases a year (8) represents a mortality rate of 0.22/100,000 (based on population estimates for 1999). Importantly, 80% of pond incidents occurred in other people's gardens.

It has been assumed that a mortality rate of 0.11/100,000 (for children under the age of five), based on 50% of that seen for garden ponds, applies to constructed wetlands as these are likely to be flooded with water much of the time, but it is felt that access would be more difficult and that self-rescue is more likely from a wetland than a pond. In addition, severe neurological handicap as a result of near-drowning (again for children under the age of five) has been assumed to occur at a rate of 0.0066/100,000 (based on 6% of the mortality figure). It has been assumed that for underground tanks in the garden there is a negligible risk of drowning or near-drowning and this has not been quantified.

5.2.2.2 *Injury*

The scope for injury in relation to greywater reuse is likely to relate to behaviour during maintenance. The system contains pumps, which could (in theory) if handled give an electric shock, although it is assumed that systems will include a residual current device (RCD), which detects some electrical faults and disconnects the supply before harm results.

An investigation of UK data suggests that pump-related electrocutions are rare. In 2005, there were 27 deaths related to exposure to electric current (other than transmission lines), these all occurred in adults and were due to all causes (National Statistics 2006), giving a rate of 0.6/million population. A study of mortality data from 1995 reported 18 deaths as a result of shocks or burns from electric current (DTI 1999) and predicted that death rates would fall by over 50% by 2010 due to the increased use of RCDs and general improvements in equipment design, despite the use of increasing numbers of electrical devices.

An examination of the RoSPA held database on injury data (Home Accidents Surveillance System), revealed 50 cases of injuries sustained through electric shock in the garden over a three year period (2000 – 2002). The majority of these (30), however, related to incidents involving hedge-trimmers or lawn mowers. Only one incident related to a pump, where the cable to a garden pond pump was inadvertently cut while the pond was being cleaned.

An additional cause of injury with a constructed wetland could be cuts obtained while maintaining the reedbed; these are likely to be small in number and trivial and have not been considered further.

Fatal electrocutions from all causes are rare in the UK. It has therefore been assumed that greywater systems will not result in any fatal electrocutions. The incidence of even minor shocks is likely to be very low (especially with the use of RCDs) and has been assumed to occur at a rate of 0.0146/100,000 in the adult population (based on figures from RoSPA).

5.2.2.3 *Chemical intoxication*

In the monitoring of five houses fitted with in-house greywater systems, Hills *et al.* (2003) found that there was a lack of control of the disinfectant dosing systems. The manufacturer's recommendation was for a total bromine level of 3–5 mg/l, but the average level recorded during the study was 40 mg/l, with at least one episode where the level exceeded 100 mg/l. The authors noted that if anti-scalant chemicals were used in the toilet bowls with these levels of total bromine there is the potential for a release of toxic gas (i.e. elemental bromine). Bromine has a strong unpleasant odour and irritates the eyes and throat and may, in severe cases, cause lung damage. This is unlikely to be an issue where the greywater is intended for external use only and has not been considered further. In addition, disinfection could result in the presence of volatile disinfection by-products (DBP), some of which are known to be respiratory irritants but, again, as the intended use is external to the household DBP formation is unlikely to result in a health effect.

Bleach and disinfectant incidents involving both adults and children are a reasonably common occurrence. Between 2000 and 2002, there were on average an estimated 4500 cases a year, with the majority of these occurring in children under the age of five. Unintentional injury involving adults and older children typically relate to incidents while using the cleaning product. Incidents in younger children generally involve ingestion of the product. Incidence rates by age group are shown in Table 5.1 (RoSPA pers.com.).

Table 5.1. Incidence rates of unintentional injury involving bleach and disinfectant

Age group	Estimated number of cases*	Incidence/100,000 population
0 to 4	2642	73.3
5 to 9	212	5.5
10 to 14	106	2.8
15 to 19	100	2.8
20 to 64	1246	3.6
65+	193	2.1

* Based on the Home and Leisure Accident Surveillance Databases (HASS and LASS)

The outcome of such unintentional injury, using domestic products, is usually very minor, whether due to ingestion of disinfectant or inhalation of fumes (Racioppi *et al.* 1994). Ingestion of small volumes (as typically occurs) usually causes minimal health effects, although in some cases it can cause burns to the mouth, throat, oesophagus and stomach and may also cause gastrointestinal irritation, nausea and vomiting (HPA 2007a). A study of 19 children in the USA, who ingested household bleach revealed no short or long-term sequelae (Harley and Collins 1997). It has been assumed that the presence of additional disinfectant in the home increases the risk of this type of incident, with a incidence rate attributable to greywater reuse of one tenth of the figures outlined in Table 5.1. Little information is available on the chlorine blocks used in greywater treatment (nor on whether children would be less likely to ingest a solid product), so it has been assumed that the severity rate and duration of disability are similar to those in the studies outlined above (i.e. mild and short lived). Thus, a severity weight 0.079 and a duration of 3 days of has been assumed for children under the age of five, while a severity weight of 0.04, with a duration of 1 day has been used for older children and adults.

5.2.2.4 *Anxiety*

Greywater reuse is currently not widely practiced in the UK. It is possible that some people could have concerns about the system. Concerns could arise as a result of:

- soil deposits;
- noxious smells;
- maintenance of an unfamiliar system; and/or;
- perceived health effects.

In single household systems, however, people do have the advantage of knowing that the reused water is a product which the household members themselves have generated. This seems to increase the acceptability as illustrated by Jeffrey and Jefferson (2003), who reported that 88% of survey respondents were willing to reuse water from their own bath/shower, compared to only 49% of respondents who were prepared to use a product derived from the whole neighbourhood. Christova-Boal *et al.* (1996) found that Australian social survey participants showed a strong preference for using greywater on gardens while, in the USA, a study found that 97% of the public surveyed approved of the idea of using re-used water for garden watering (Lohman 1987).

Anxiety qualitative estimate

When the systems are working correctly with a fail-safe design which ensures that greywater is delivered to the sewer if there is a blockage, a treatment failure or an extended period of storage, noxious smells should be minimised. Some people may have concerns about the maintenance and cleaning of filters but, as each system is serving a single household, this is likely to be no more unpleasant than removing hair from a shower plug or cleaning a domestic toilet. Although anxiety may be a real issue, especially in the elderly (in relation to perceived maintenance requirements), it is currently not possible to quantify and should be minimised in single household systems with a maintenance contract, thus a qualitative estimate of (−) is suggested to indicate a possible negative impact.

5.2.2.5 *Infection*

Infection relating to greywater reuse could occur via inappropriate ingestion/contact; inhalation; microbial contamination of the environment and vector-borne illness. Ingestion could occur via a number of routes, such as:

- ingestion of aerosols produced through car washing or garden watering;

- direct ingestion via the garden tap; and/or
- ingestion of garden produce and/or soil watered with greywater.

Contact could occur during car washing, garden watering, filter maintenance and, perhaps, the inappropriate use of greywater for filling paddling pools, hot tubs and so on. Inhalation of pathogens could occur as a result of aerosols generated during car washing or garden watering. Vector-borne illness could occur as a result of stored water being a potential breeding site for mosquitoes or as a result of the greywater system encouraging an increase in vermin, such as rats.

Pathogens in greywater are derived principally from human faecal material (although where, for example, pets are washed in the bath animal faecal material may also be present). To a lesser extent, pathogens may also be present in urine and be derived from the skin and nose and throat secretions (e.g. *Staphylococcus aureus*). Clearly, however, while raw greywater may always be contaminated with faecal material, faecal pathogens will only be present when a household member has an infection.

Examination of water quality data (section 5.1.1) revealed that a number of human pathogens have been isolated from greywater, including *Pseudomonas aeruginosa* and *Cryptosporidium* spp. although, in theory, any pathogen excreted from the body could be present. A number of reference pathogens (consisting of representatives from each of the major pathogen groups – i.e. bacteria, protozoa and viruses) have been investigated to determine the potential for infection risks.

Bacteria

Although *Campylobacter* spp. is the leading cause of bacterial enteritis (stomach upset) in the UK, it is very susceptible to disinfection. So a more robust bacterial pathogen such as *Salmonella* spp. provides a more appropriate (precautionary) choice (although levels of this pathogen will also be reduced by disinfection). In England and Wales, two types of *Salmonella* spp. (*S. typhimurium* and *S. enteritidis*) typically account for over 75% of reported cases. It has been estimated that the annual incidence rate is 218 cases/100,000 population (Adak *et al.* 2002).

Protozoa

Cryptosporidium spp. is a robust protozoan parasite which is resistant to disinfection and has been identified in greywater (Birks *et al.* 2004). Reports of illness in England and Wales average over 4000 per year (HPA 2006a). This has been estimated to result in 42,000 actual annual cases of illness (i.e. accounting for those cases where people do not go to the doctor or no pathogen is isolated), which is an incidence rate of 79/100,000 (Adak *et al.* 2002).

Viruses

Norovirus and rotavirus are the most common cause of viral gastroenteritis in the UK, with estimates of incidence of 12.5 cases/1000 population and 7.1 cases/1000 population respectively (Adak *et al.* 2002). Both are also known to be shed in the faeces in extremely high numbers (e.g. up to 10^{11} per gram of faeces – Rusin *et al.* 2000). Currently, however, there is no dose-response relationship for norovirus, which limits its use in quantitative microbial risk assessment (QMRA), therefore rotavirus has been chosen as the reference pathogen. Person-to-person transmission and inhalation of airborne virus (or aerosols containing virus) are thought to be the key routes of infection. In temperate climates, however, infections tend to be seasonal and mostly occur in the winter months (see Figure 5.1), when greywater reuse (in the current scenario) is likely to be minimal. Rotavirus transmission via greywater has not, therefore, been considered further here. It will, however, be important to estimate possible infection where greywater is used for toilet flushing (a year round activity).

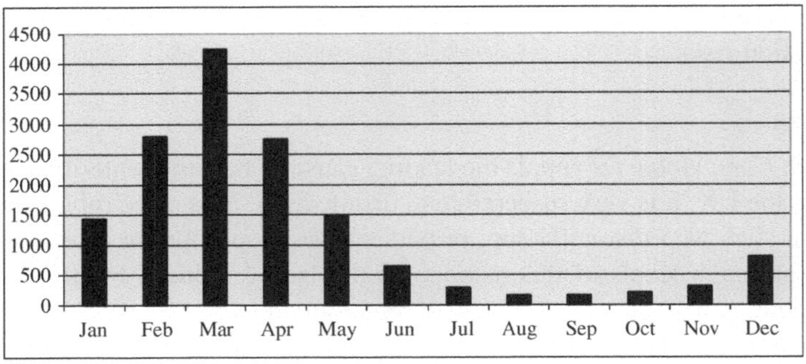

Figure 5.1. Rotavirus infection by month

In addition to the reference enteric pathogens chosen, it is likely that a number of non-enteric bacterial pathogens, such as *Ps. aeruginosa* and *Legionella* spp., could be present in greywater.

Although nutrient levels in greywater tanks may be sufficient to allow mosquito development, properly installed underground tanks should not permit their access. The fluctuating water levels within the constructed wetland should prevent mosquito breeding. Thus, greywater reuse systems should not lead to a change in mosquito levels.

There is a small possibility that constructed wetlands could lead to an increase in vermin, such as rats and mice. Rats are known to excrete pathogenic *Leptospira* spp. in their urine. In damp conditions these may survive and could present a hazard where people are maintaining their wetland (despite the maintenance contract), especially if they have, or sustain, any cuts on their hands and arms (transmission is via wound infection). It is not known how, or if, the presence of individual wetlands will impact on vermin numbers. This coupled with the low incidence of leptospirosis in the UK (only 37 cases were reported in England and Wales during 2004 [CDR 2005], representing an incidence rate of 0.07/100,000 population) suggest that the hazard will probably be negligible. However, if climate change leads to an increase in the incidence of leptospirosis (it is more common in warmer countries) and there is an increase in vermin (either as a result of climate change or the use of individual constructed wetlands) this hazard should be reviewed.

The risk of infection presented by greywater reuse (in the defined scenario) was investigated by looking at the following possibilities: (appropriate elements of which were subjected to QMRA)

- ingestion of *Salmonella* spp, from aerosol formed during hose use, e.g. car washing;
- ingestion of *Salmonella* spp. and *Cryptosporidium* spp. due to direct ingestion of greywater, via the garden tap;
- ingestion of garden produce contaminated with *Cryptosporidium* spp.;
- ingestion of *Cryptosporidium* spp. directly from soil;
- inhalation of *Legionella* spp. from aerosol formed during hose use; and
- skin contact with greywater used to fill a paddling pool.

5.2.2.5.1 Salmonella spp.

Salmonella spp. is a bacterial pathogen capable of causing human infection. It is the second most commonly reported aetiological agent of bacterial gastroenteritis in England and Wales. Salmonellosis generally presents as diarrhoea and abdominal pain, sometimes with fever and vomiting. In a small number of cases, hospitalization may be required.

The dose-response curve for salmonellosis is based on pooled data for a number of *Salmonella* spp. isolated from eggs, tested on healthy volunteers

(Haas *et al.* 1999). The Beta-Poisson model was found to provide the best fit of the data, with constants of $\alpha = 0.3126$ and β=2884. Teunis *et al.* (1999) report a dose-response relationship of $\alpha = 0.89$ and $\beta = 4.4 \times 10^5$ for *S. meleagridis* based on the same human volunteer data. It should be noted that none of the *Salmonella* strains examined were those associated with most salmonellosis in the UK, i.e. *S. typhimurium* and *S. enteritidis* (HPA 2006b).

It has been assumed that 50% of those infected with *Salmonella* will develop clinical symptoms (Rusin *et al.* 2000) that last between 4 to 7 days (CDC 2007). The severity weight of 0.093 is based on an uncomplicated episode of diarrhoea, while cases requiring hospitalization (1.16% of those who become ill – Mead *et al.* 1999) have been attributed a severity weight of 0.42 (VGDHS 1999). In the UK, the majority of fatalities from intestinal infectious diseases occur in those aged 65+ (National Statistics 2003). Therefore, the age at death in the case of fatal infections (0.041% of those who become ill – Mead *et al.* 1999) is assumed to be 78, the same as that for campylobacteriosis (Havelaar *et al.* 2000a).

Exposure to Salmonella spp. through ingestion of water during hose activities

The data and assumptions made in estimating the exposure to *Salmonella* spp. through the use of treated greywater delivered by hose and associated devices are shown in Table 5.2.

5.2.2.5.2 Cryptosporidium spp.

Cryptosporidium spp. is a protozoan pathogen capable of causing human infection. Cryptosporidiosis is usually self-limiting diarrhoea which can sometimes include nausea, vomiting and fever. Although it usually resolves within a week in otherwise healthy people, it can last for a month or more. The severity of illness depends on age and immune status; infections in immunocompromised people can be severe and even life-threatening. The most commonly used dose-response relationship is an exponential model (Teunis *et al.* 1996; WHO 2002). A duration of 6 days, with a severity weight of 0.054 has been assumed for the immunocompetent ('normal') population (with 71% of infections progressing to illness), while a duration of 47 days and a severity weighting of 0.13 has been assumed for AIDS patients, where all infection progresses to illness (Havelaar *et al.* 2000b). As almost 90% of the case study population classified their health as good a rate of 0.05% HIV/AIDS prevalence

Table 5.2. Exposure to *Salmonella* spp. through hose activities (ingestion)

Data requirements	Comments
The frequency of *Salmonella* spp. infection	Unlike faecal indicator organisms, people do not constantly excrete pathogens. There is little, if any, data on pathogen concentrations in raw greywater (although some studies have reported pathogen presence or absence). It has been estimated that the annual incidence of illness due to salmonellosis in the UK is 218/100,000 population (Adak *et al.* 2002). Cases, however, have a seasonal basis towards the summer months (HPA 2007b). Frequency of infection was calculated by using monthly percentage reporting figures and applying them to the annual incidence to derive a monthly incidence.
Level of *Salmonella* spp. excretion	Excretion of *Salmonella* has been estimated to range between 10^4 and 10^{10} organisms per gram of faeces (Rusin *et al.* 2000) and last between three and five days
Greywater production	Metered households have an average per capita consumption (pcc) of drinking water of 139 l/head/day (OFWAT 2005). It has been assumed that with a greywater system, approx. 30% of this would become greywater. Based on occupancy figures for the case study it is estimated that daily greywater production is 105 l per household.
Greywater contamination	Ottoson and Stenström (2003a) concluded that the best measure of faecal load in greywater was provided by intestinal enterococci, which resulted in an estimate of 5.4 g/person/day. It was assumed that when a householder is ill they will contaminate the greywater with this level of faecal material containing excreted pathogens (at the level described above).
Microbial removal efficiency (wetland)	Based on results from Cranfield University, a vertical flow constructed wetland resulted in an average $\log_{10}$ reduction of between 3 and 4.75 for faecal indicator bacteria (Jefferson, pers. com.). These results are in line with figures reported elsewhere (e.g. WHO 2006a).

Table 5.2. *Continued*

Data requirements	Comments
Microbial removal efficiency (chlorine disinfection)	Diaper *et al.* (2001), used a 4 $\log_{10}$ reduction for chlorine disinfection alone and an assumption that salmonellae "*behave in a similar manner to faecal coliforms in a greywater system*". As the greywater in the current scenario will already have received some treatment a lower log reduction of between 2 and 3 has been assumed.
The fraction of *Salmonella* spp. organisms capable of initiating infection	Dose-response model of Haas *et al.* 1999, where $\alpha = 0.3126$ and $\beta = 2884$
Hose use	It has been assumed that hoses utilising treated greywater will be used for a number of purposes, i.e. car washing, garden watering and play. Usage for car washing has been based on figures from Herrington (1996), who estimated that 15% of households wash their cars using a hose twenty times a year, using 150 litres of water. It has been assumed that this occurs throughout the year. Garden watering using a hose or sprinkler was assumed to occur between May and September. During June to August it has been assumed that children (above the age of four) will incorporate the garden watering into their play. 50% (range 40 – 60) of households were assumed to water their garden in this way.
Ingestion volume	During car washing and general garden watering it has been assumed that only one person is exposed and that they ingest a maximum of 1ml on each occasion. Dufour *et al.* 2006 examined the ingestion of water during swimming activities (where there is far greater potential for ingestion) and found that on average children ingested 37 ml of water, over a 45 minute period (range 0 to 154 ml). So, when children are playing it has been assumed that they could ingest a maximum of 15ml per exposure, based on a tenth of the max figure reported by Dufour.

has been assumed, rather than the estimated UK national level of 0.2% (CIA 2005).

Exposure to Salmonella spp. and Cryptosporidium spp. through direct ingestion (external tap use)

The data and assumptions required to estimate the exposure of children to *Salmonella* spp. through direct ingestion, via the garden tap, is shown in Table 5.3, while that for *Cryptosporidium* spp. is in Table 5.4.

Table 5.3. Exposure of children to *Salmonella* spp. through direct ingestion (via the garden tap)

Data requirements	Comments
The frequency of *Salmonella* spp. infection	See Table 5.2
Level of *Salmonella* spp. excretion	See Table 5.2
Greywater production	See Table 5.2
Greywater contamination	See Table 5.2
Microbial removal efficiency	See Table 5.2
The fraction of *Salmonella* spp. capable of initiating infection	See Table 5.2
The amount of un-boiled water consumed	A UK report on water consumption (MEL, 1996) reported that the average amount of tap water consumed daily was 1138ml of which un-boiled water accounted for 16.7% (or 190ml)
The number of children likely to drink from external taps	It has been assumed that during the 6 week summer holiday period, between 0 and 5% of children aged between 5 and 14 drink from the outside tap on an occasional basis (between 0.5 and 2 times a week) and consume between 50 and 250ml on each occasion

Table 5.4. Exposure of children to *Cryptosporidium* spp. through direct ingestion (via the garden tap)

Data requirements	Comments
The frequency of *Cryptosporidium* spp. infection	It has been estimated that the annual incidence of cryptosporidiosis in the UK is 79/100,000 (Adak *et al.* 2002). There is a seasonal variation in reporting, and based on this data it has been calculated that the summer incidence rate is 30.5/100,000.
Level of *Cryptosporidium* spp. excretion	The shedding of oocysts in faeces during infection is typically between 10^6 and 10^7 organisms per gram of faeces (Rusin *et al.* 2000) and typically lasts for up to 14 days (Chappell *et al.* 1996)
Greywater production	See Table 5.2
Greywater contamination	See Table 5.2
Microbial removal efficiency	This has been assumed to be similar to that for *Salmonella* spp., with an overall $\log_{10}$ reduction of between 3 and 6
The fraction of *Cryptosporidium* capable of initiating infection	The dose-response is represented by an exponential model where r = 0.004005
The amount of un-boiled water consumed	See Table 5.3
The number of children likely to drink from external taps	See Table 5.3

Exposure to Cryptosporidium spp. through contaminated garden produce

Where treated greywater is used to water garden produce, such as salad crops and foodstuffs, there is the potential for ingestion of pathogens. The data and assumptions required to investigate exposure via this route are outlined in Table 5.5.

No account was taken of pathogen removal with washing, or of people other than the householders growing the produce eating contaminated salad crops (i.e. if householders give away produce).

Exposure to Cryptosporidium spp. through soil ingestion

Incidental soil ingestion occurs at all ages, through poorly washed food and as a result of both hand- and object-to-mouth activities. It is however, particularly marked in young children as a result of their mouthing behaviour (Cohen Hubal

Table 5.5. Exposure to *Cryptosporidium* spp. through the ingestion of contaminated garden produce

Data requirements	Comments
The frequency of *Cryptosporidium* spp. infection	See Table 5.4
Level of *Cryptosporidium* spp. excretion	See Table 5.4
Greywater production	See Table 5.2
Greywater contamination	See Table 5.2
Microbial removal efficiency	See Table 5.4
The fraction of *Cryptosporidium* spp. capable of initiating infection	See Table 5.4
Number of households growing salad plants (all of which are assumed to be watered with treated greywater)	A MORI poll, commissioned by the Royal Horticultural Society suggests that 41% of people grow their own fruit and vegetables and that over a two-year period 12% of these people have attempted to grow lettuce (RHS 2004; MORI 2004). Thus it has been assumed that each year 6% of the fruit and vegetable growing population (410/1000) grow lettuce (i.e. 24.6/1000 population).
Consumption of home grown salad plants	Figures from Northern Ireland (NFS 2005) suggest an average weekly lettuce (and leafy salad) consumption of 25.9 g/person. It has been assumed that families who grow their own vegetables will eat double this amount of their own produce between June and October, i.e. during a five month season.
Level of salad crop contamination and pathogen retention	Hamilton *et al.* (2006) used a mean value of 0.108 ml/gram of lettuce (with a standard deviation of 0.019, with the distribution truncated at 0) – based on the data from Shuval *et al.* 1997

et al. 2000). Mouthing is an important component of childhood development and is an exploratory behaviour where objects are placed into the mouth briefly for purposes of discovery. During this stage of development, children will put their hands and any object they come into contact with into their mouths (Tulve *et al.* 2002). Mouthing behaviour is most common in children aged 24 months or less

(Tulve *et al.* 2002), where mouthing frequency is roughly double (81 events an hour) that seen in children older than 24 months (42 events an hour).

In addition to inadvertent ingestion of soil, many children also actively swallow soil. Davis and Mirick (2006) examined the incidence of geophagy (eating of soil) in a group of 19 children in the USA and found that almost 60% of the children swallowed dirt at least once a week. Children's mean soil ingestion ranged from 37 to 207 mg/day, depending upon the tracer used (Davis and Mirick 2006).

Clearly, when there are a relatively small number of cases of illness in the population (i.e. approximately 4 cases per year), this implies that exposure is relatively unlikely – i.e. a number of factors would need to coincide in order for exposure to occur. This is especially the case in this example where it is only really likely that children less than 5 will be affected. It is estimated from the census figures that 1 in 4 households have a child this age. Assuming that cryptosporidiosis and the presence of a young child in the household are not related, it might be expected that only 1 child could be exposed. As soil ingestion is more likely when the weather is good this further reduces the likelihood of exposure as the case of illness would have to be during the summer months when garden watering was happening. The following, rather pessimistic, assumptions have therefore been used to explore two soil consumption scenarios:

- infection in the household coincides with irrigation and child-related ingestion of contaminated soil;
- children in affected households eat soil on two occasions when ground is contaminated (consuming either 200mg each time or 50 mg each time from the contaminated area);
- organisms per ml are assumed to translate to organisms/gram of soil;
- there is no reduction in organism numbers between watering and ingestion;
- watering takes place once during the contamination incident;
- four households (expected number 2.2) have a case of cryptosporidiosis during the summer (when children are most likely to be outside and be exposed to soil); each of these households has a child under the age of five who regularly eats soil (who is not the primary case); and
- treatment efficiency is at the low end of the range (i.e. 3 log removal for the VFCW).

5.2.2.5.3 *Legionella* spp.

Legionella spp. are respiratory pathogens with *Legionella pneumophila*, the cause of legionnaires' disease, being the main pathogen of concern. Transmission is through the inhalation of contaminated aerosols. Disease caused by *L. pneumophila* has two distinct clinical presentations: Legionnaires' disease (a form of pneumonia with a low attack rate and a relatively high fatality rate) and Pontiac fever (an acute, self-limiting illness with flu-like symptoms), collectively known as legionellosis.

Legionella thrive in biofilms; because bacteria in biofilms are relatively resistant to standard water disinfection procedures, legionellae are able to enter and colonize potable water supplies (EPA 2001). Indeed, showers (including domestic showers) have been occasionally implicated in the transmission of Legionnaires' disease (Castellano Pastoris *et al.* 1988) and domestic hot water systems and showers have been shown to harbour *Legionella* spp. (e.g. Borella *et al.* 2004). In one study in Canada, over 30% of houses (69 from a total of 211) tested had at least one sample (i.e. taken from either the water heater or the most frequently used tap or shower) positive for *Legionella* spp. (Alary and Joly 1991). In Italy, *Legionella* spp. were detected in 22% of samples drawn from bathroom outlets, with a geometric mean concentration of 1.17×10^3 cfu/l (Borella *et al.* 2004). Although it has been noted that the risk of infection following exposure to *Legionella* is difficult to assess and remains a matter of some debate (Atlas 1999).

The presence of legionellae in hot water systems means that they may, in some instances, be present in greywater tanks, where there may be potential for growth. *Legionella* populations are known to be enhanced by heat and elevated temperatures (EPA 2001) and can exhibit growth at temperatures above 25 °C, temperatures which have been occasionally recorded for stored greywater in the UK (Winward, pers. comm.).

Qualitative estimate of legionellosis

No information has been found in the literature on concentrations of *L. pneumophila* in greywater. This, in conjunction with the fact that there is currently no dose-response relationship relating to infection by *L. pneumophila* in humans, means that it is not possible to perform even an exploratory microbial risk assessment. However, even if legionellae are present in treated greywater and are capable of growth, it seems unlikely that exposure through hose use would be greater than that through showering. Thus, a qualitative estimate of (−) has been assigned to this hazard

5.2.2.5.4 Skin infection

There are a number of opportunistic pathogens which may be present in raw greywater which can cause skin infections following prolonged contact. These include *Pseudomonas aeruginosa*, *Mycobacterium* spp. and *Staphylococcus aureus*. These bacteria have all been associated with skin rashes in swimming pool environments (WHO 2006b).

Greywater treatment would be expected to reduce these bacteria down to very low levels. If, however, a paddling pool is filled and left for some time it is possible that they may proliferate.

Qualitative estimate of skin infection

There are no data on the levels of these dermal bacteria in treated greywater, possible dose-response or likely levels of proliferation so it is not possible to conduct a microbial risk assessment, although it would be expected (based on case reports from swimming pools) that health impacts would be uncommon and mild (Yoder *et al.* 2004), resulting in a qualitative estimate of (−). This may not, however, be the case if treated greywater were used to fill a garden hot tub, where warm temperatures and water agitation make it likely that high levels of proliferation could be seen, and it is strongly suggested that greywater is **not** used for this purpose.

5.2.3 Risk characterization

The estimates relating to drowning are determined by multiplying the estimated number of cases by the amount of years of life lost (assuming an average age at death of 2.5 years and a UK average age at death of 79). Estimates for neurological handicap and injury were determined by multiplying the number of occurrences by the severity weight and duration to calculate the years lived with a disability.

5.2.3.1 Drowning and near-drowning

The analysis of exposure (section 5.2.2.1) estimated that 0.0005 children could drown each year as a result of the garden-based constructed wetlands, and that a further 0.00003 could suffer severe neurological handicap (severity weight 0.76) as a result of near-drowning. These estimates result in DALY scores of 0.039 and 0.0018 respectively (see Table 5.6).

5.2.3.2 Injury and chemical intoxication

The assumed rate of pump-related electrocutions results in an annual DALY score of 2.2×10^{-6}. Disinfectant-related injuries were estimated to occur in 0.0036 children and 0.0013 adults.

Table 5.6. Risk characterization summary (greywater reuse)

Hazard	Exposure	Group	Estimate	Annual cases	DALYs
Drowning		Children < 5		5.2×10^{-4}	0.039
Near-drowning		Children < 5		3.1×10^{-5}	0.0018
Injury	Pump electrocutions				
		Adults		5.1×10^{-4}	2.24×10^{-6}
Chemical intoxication		Children		3.6×10^{-3}	2.25×10^{-6}
		Adults		1.3×10^{-3}	1.42×10^{-7}
Mental health – anxiety			Qual		(−)
Illness	Salmonellosis from hose activities				
		All	Min	4.4×10^{-5}	9.5×10^{-8}
			Mean	1.1×10^{-2}	2.3×10^{-5}
			Max	0.051	1.1×10^{-4}
	Salmonellosis from direct ingestion (external tap)				
		Children	Min	0	0
			Mean	0.003	6.2×10^{-6}
			Max	0.062	1.3×10^{-4}
	Cryptosporidiosis from direct ingestion (external tap)				
		Children	Min	0	0
			Mean	0.022	1.9×10^{-5}
			Max	0.272	2.4×10^{-4}
	Cryptosporidiosis from ingestion of garden produce				
		Norm pop	Min	8×10^{-6}	8.4×10^{-9}
			Mean	0.022	2.2×10^{-5}
			Max	0.266	2.6×10^{-4}
		Immunocomp	Min	4.6×10^{-9}	7.7×10^{-11}
			Mean	1.2×10^{-5}	2.0×10^{-7}
			Max	1.4×10^{-4}	2.4×10^{-6}
	Cryptosporidiosis from soil ingestion				
		Children	Worst case∗	0.27	2.4×10^{-4}
	Legionellosis				
			Qual		(−)
	Pseudomonas infection				
			Qual		(−)
	Vector-borne illness				
		All		0	0

Norm pop: Normal population
Immunocomp: Immunocompromised population
Min: minimum estimate
Mean: mean estimate
Max: maximum estimate
Qual: qualitative estimate
(−): minor health impact
∗ Worse case mean estimate

5.2.3.3 Infection

Where possible, parameters for the individual infection exposures were entered into @Risk (Palisade Corporation 2002) as probability distributions rather than point estimates, in order to examine the effects of uncertainty and the assumptions to which the estimates were most sensitive. The data for each of the infection exposures are summarised in Appendix A5.1.

5.2.3.4 Risk characterization summary

Table 5.6 summarises the risk characterization for each of the identified hazards applied to the case study population.

5.2.4 Risk management

The most obvious risk management strategy is to ensure that all garden-based constructed wetlands are fenced in an appropriate manner to prevent access for children under the age of five. This is important, irrespective of whether the family has children that age or not as studies have shown that up to 80% of pond incidents occurred in other people's gardens (see section 5.2.2.1). Where garden swimming pools have been appropriately fenced, drownings and near-drownings in the under fives have been cut dramatically (Thompson and Rivara 2000).

It may be possible to advise homeowners to divert their greywater to waste when someone in the household has a gastrointestinal illness. However, given the relatively low levels of infection predicted to occur as a result of using either of the systems this may not be a desirable option. Some pathogens may be shed before and after obvious gastrointestinal symptoms have ceased, so exposure may still occur and it may raise perception problems about the systems which decrease public confidence unnecessarily.

Measures to preclude the use of external taps for provision of drinking water would further reduce infection risk, although it should be noted that the assumed number of children using garden taps is very speculative and additional research might be useful to determine whether this is an actual issue prior to making the use of external taps more difficult (as any measure which makes the system more difficult to use is likely to reduce water savings).

Household manuals detailing the greywater recycling systems should stress simple hygiene measures, such as covering exposed cuts is working on the wetland, and wearing gloves when cleaning the coarse filters.

5.3 HEALTH IMPACT STATEMENT

The annual DALY score for constructed wetlands is 4.1×10^{-2}. Much of this score (3.9×10^{-2}) is attributable to the predicted effects due to drowning. Restricting physical access, by fencing the wetlands or, perhaps, raising them, to ensure that children under the age of five can not access them dramatically reduces the DALY score.

Total DALY score (with mitigation) 3.1×10^{-4}
Qualitative estimate: 3 (–)

5.4 DISCUSSION

The success of a greywater reuse system, both in terms of minimising health impacts and maximising water savings, will depend upon the system being properly maintained. This HIA has assumed that this is the case, although it has used a range of treatment efficiencies which will capture (at least to some degree) periodic poor performance. While the HIA has assumed that the systems are working correctly it has used a number of precautionary (i.e. probably pessimistic) assumptions, which serve to maximise the predicted contamination, exposure and hence projected infection. Typically, for example, it has been assumed that contamination and exposure coincide.

In some ways, the scenario presented here (i.e. the use of greywater solely for garden watering and external use) is not a realistic one, as there will be little or no use of the greywater during winter. Clearly, this is unrealistic if water savings are to be maximized. Additionally, the scenario is deliberately lax in terms of safeguards (i.e. the treated greywater is available for car washing, which could result in aerosol production) to facilitate exploration of possible risks and suggest mitigation measures as appropriate. The results suggest that the infection risks are low, although not all pathogens and possible illnesses have been considered. For individual household systems, the use of treated greywater for hose use represents an additional route of exposure to pathogens when someone within the household is ill. Compared to person-to-person contact, and other more common routes of secondary transmission it will (except in very unusual cases) be an extremely minor route of infection.

Restricting the access of small children to the constructed wetlands means that the estimated risk from greywater reuse in the scenario described falls well within both the screening level and WHO reference level of risk (Chapter 2), as shown in Figure 5.2, where risks for the case study population are compared on a $\log_{10}$ scale.

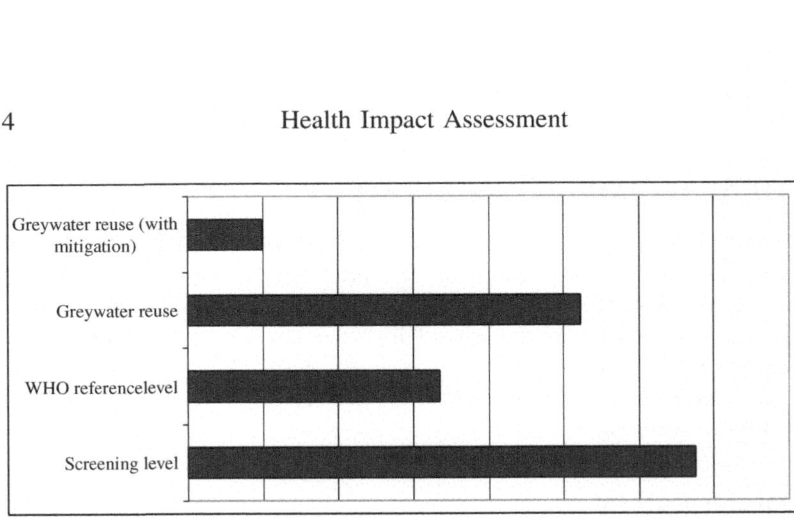

Figure 5.2. Comparison of case study population risks ($\log_{10}$ scale)

5.5 REFERENCES

Adak, G.K., Long, S.M. and O'Brien, S.J. (2002) Trends in indigenous foodborne disease and deaths, England and Wales: 1992 to 2000. *Gut* **51**, 832–841.

Alary, M. and Joly, J.R. (1991) Risk factors for contamination of domestic hot water systems by Legionellae. *Applied and Environmental Microbiology* **57**(8), 2360–2367.

Atlas, R.M. (1999) *Legionella*: from environmental habitats to disease pathology, detection and control. *Environmental Microbiology* **1**(4), 283–293.

Birks, R., Colbourne, J., Hills, S. and Hobson, R. (2004) Microbiological water quality in a large in-building, water recycling facility. *Water Science and Technology* **50**(2), 165–172.

Borella, P., Montagna, M.T,, Romano-Spica, V., Stampi, S., Stancanelli, G., Triassi, M., Neglia, R., Marchesi, R., Fantuzzi, G., Tato, D., Napoli, C., Quaranta, G., Laurenti, P., Leoni, E., De Luca, G., Ossi, C., Moro, M. and Ribera D'Alcala, G. (2004) Legionella infection risk from domestic hot water. *Emerging Infectious Diseases* **10**(3), 457–464.

Brenner, R.A. (2002) Childhood drowning is a global concern. *British Medical Journal* **324**, 1049–1050.

Burrows, W.D., Schmidt, M.O., Carnevale, R.M. and Schaub, S.A. (1991) Non potable reuse: Development of health criteria and technologies for shower water recycle. *Water Science and Technology* **24**(9), 81–88.

Castellani Pastoris, M., Vigano, E.F. and Passi, C. (1988) A family cluster of *Legionella Pneumophila* infections. *Scandinavian Journal of Infectious Disease* **20**(5), 489–493.

CDC (2007) Salmonellosis. http://www.cdc.gov/ncidod/dbmd/diseaseinfo/salmonellosis_g.htm (accessed 20/04/07).

CDR (2005) Common animal associated infections, England and Wales laboratory reports: weeks 49-52/04. *Communicable Disease Report Weekly* **15**(6).

Chappell, C.L., Okhuysen, P.C., Sterling, C.R. and DuPont, H.L. (1996) *Cryptosporidium parvum*: Intensity of infection and oocyst excretion patterns in healthy volunteers. *Journal of Infectious Diseases* **173**, 232–236.

Christova-Boal, D., Eden, R.E. and McFarlane, S. (1996) An investigation into greywater reuse for urban residential properties. *Desalination* **106**, 391–397.

CIA (2005) The World Fact Book 2005. Central Intelligence Agency, USA. http://www.cia.gov/cia/publications/factbook/geos/uk.html (accessed 07/12/05).

Cohen Hubal, E.A., Sheldon, L.S., Burke, J.M., McCurdy, T.R., Berry, M.R., Rigas, M.L., Zantarian, V.G. and Freeman, N.C.G. (2000) Children's exposure assessment: a review of factors influencing children's exposure, and the data available to characterize and assess that exposure. *Environmental Health Perspectives* **108**(6), 475–486.

Davis, S. and Mirick, D.K. (2006) Soil ingestion in children and adults in the same family. *Journal of Exposure Science and Environmental Epidemiology* **16**(1), 63–75.

Diaper, C., Dixon, A., Butler, D., Fewkes, A., Parsons, S.A., Strathern, M., Stephenson, T. and Strutt, J. (2001) Small scale water recycling systems – risk assessment and modelling. *Water Science and Technology* **43**(10), 83–90.

Dixon, A., Butler, D., Fewkes, A. and Robinson, M. (1999a) Measurement and modelling of quality changes in stored untreated greywater. *Urban Water* **1**, 293–306.

Dixon, A., Butler, D. and Fewkes, A. (1999b) Water saving potential of domestic water reuse systems using greywater and rainwater in combination. *Water Science and Technology* **39**(5), 25–32.

DTI (1999) Research on the pattern and trends in home accidents. Department of Trade and Industry, London.

DTI (2000) Drowning accidents in the garden involving children under five.Department of Trade and Industry.

Dufour, A.P., Evans, O., Behymer, T.D. and Cantú, R. (2006) Water ingestion during swimming activities in a pool: a pilot study. *Journal of Water and Health* **4**(4), 425–430.

EA (2000) A study of domestic greywater recycling. National Water Demand Management Centre. Environment Agency, Worthing. pp 59. April 2000.

EPA (2001) Legionella: Drinking water health advisory. United States Environmental Protection Agency, Washington. EPA-822-B-01-005. http://www.epa.gov/waterscience/criteria/humanhealth/microbial/legionellaha.pdf (accessed 9/5/07)

Friedler, E. (2004) Quality of individual domestic greywater streams and its implication for on-site treatment and reuse possibilities. *Environmental Technology* **25**, 997–1008.

Grant, A., Gray, S., Hansen, D., Murdoch, B. and Reid, H. (2006) Sustainable water systems schemes and technologies: Monitoring results from Inkerman D'Lux. CSIRO Urban Water, Australia.

Haas, C.N., Rose, J.B. and Gerba, C.P. (1999) Quantitative Microbial Risk Assessment. Wiley, New York.

Hamilton, A.J., Stagnitti, F., Premier, R., Boland, A-M. and Hale, G. (2006) Quantitative microbial risk assessment models for consumption of raw vegetables irrigated with reclaimed water. *Applied and Environmental Microbiology* **72**(5), 3284–3290.

Harley, E.H. and Collins, M.D. (1997) Liquid household bleach ingestion in children: a retrospective review. *Laryngoscope* **107**(1), 122–125.

Havelaar, A.H., de Wit, M.A.S., van Koningsveld, R. and van Kempen, E. (2000a) Health burden in the Netherlands due to infection with thermophilic *Campylobacter* spp. *Epidemiology and Infection* **125**, 505–522.

Havelaar, A.H., Hollander de, A.E.M., Teunis, P.F.M., Evers, E.G., Kranen van, H.J., Versteegh, F.M., Koten van, J.E.M. and Slob, W. (2000b) Balancing the risks and benefits of drinking-water disinfection: disability-adjusted life years on the scale. *Environmental Health Perspectives* **108**, 315–321.

Herrington, P. (1996) Climate Change and the Demand for Water, HMSO, London

Hills, S., Birks, R., Diaper, C. and Jeffrey, P. (2003) An evaluation of single-house greywater recycling systems. *Presented at the International Water Association (IWA) Conference on Reclamation and Reuse, Mexico November 2003.*

HPA (2006a) *Cryptosporidium* Laboratory Reports – all identifications England and Wales, 1986–2005. Health Protection Agency, London.

HPA (2006b) *Salmonella* in humans (excluding *S. typhi* and *S. paratyphi*) faecal and lower gastrointestinal isolates: England and Wales, 1981 – 2005. Health Protection Agency, London. http://www.hpa.org.uk/infections/topics_az/salmonella/data_human.htm (accessed 06/12/06).

HPA (2007a) Sodium hypochlorite: human effects of acute/single exposure (human). Health Protection Agency, London. http://www.hpa.org uk/Chemicals/compendium/sodium%20hypochlorite/acute.htm (accessed 16/04/07).

HPA (2007b) *Salmonella* in humans (excluding *S. typhi* and *S. paratyphi*) by month. England and Wales, 1992 – 2006. http://www.hpa.org.uk/infections/topics_az/salmonella/data_human_month.htm (accessed 24/04/07)

Jeffrey, P. and Jefferson, B. (2003) Public receptivity regarding 'in-house' water recycling: results from a UK survey. *Water Science and Technology: Water Supply* **3**(3), 109–116.

Kemp, A. and Sibert, J.R. (1992) Drowning and near drowning in children in the United Kingdom: lessons for prevention. *British Medical Journal* **304**, 1143–1146.

Leggett, D.J., Brown, R., Stanfield, G., Brewer, D. and Holliday, E. (2001) Rainwater and greywater use in buildings: decision-making for water conservation. CIRIA PR80. Construction Industry Research and Information Association, London.

Lohman, L. (1987) Potable wastewater re-use can win public support. In *Proceedings of Water Re-use Symposium*, August 2–7, 1987, Denver, Colorado. AWWA Research Foundation, Denver, CO.

Mead, P.S., Slutsker, L., Dietz, V., McCaig, L.F., Bresee, J.S., Shapiro, C., Griffen, P.M. and Tauxe, R.V. (1999) Food-related illness and death in the United States. *Emerging Infectious Diseases* **5**(5), 607–625.

MEL (1996) Tap water consumption in England and Wales: findings from the 1995 national survey. MEL Research Report 9448-01.

MORI (2004) Good Life Factor – RHS. Omnibus questions – topline results. MORI/22234. http://www.chelseashow.com/publications/pubs/garden0704/morireporrttopline.pdf (accessed 16/01/06).

National Statistics (2003) Mortality Statistics – Cause. Review of the Registrar General on deaths by cause, sex and age in England and Wales, 2003. Series DH2 no. 30. Office for National Statistics, London.

National Statistics (2006) Mortality statistics: cause. Review of the Registrar General on deaths by cause, sex and age, in England and Wales, 2005. Office for National Statistics, London.

NFS: Quarterly Consumption of Household Food (2005) www.dardni-gov.uk/econs/foodconsumption03_04/econ40d.xls (accessed 12/12/05)

Noah, M. (2002) Graywater use still a gray area. *Journal of Environmental Health* **64**(10), 22–25.

Nolde E. (1999) Greywater reuse systems for toilet flushing in multi-storey buildings – over ten years experience in Berlin. *Urban Water* **1**, 275–284.

OFWAT (2005) Security of Supply, Leakage and the Economic Use of Water: 2004/05 Report. OFWAT, London.

Ottoson, J. and Stenström, T.A. (2003a) Faecal contamination of greywater and associated microbial risks. *Water Research* **37,** 645–655.

Ottoson, J. and Stenström, T.A. (2003b) Growth and reduction of microorganisms in sediments collected from a greywater treatment system. *Letters in Applied Microbiology* **36**, 168–172.

Palisade Corporation (2002) Guide to using @Risk: risk analysis and simulation add in for Microsoft Excel. Version 4.5. Palisade Corporation, Newfield, New York. http://www.palisade.com accessed April 14 2008.

Palmquist, H. and Hanæus, J. (2005) Hazardous substances in separately collected grey- and blackwater from ordinary Swedish households. *Science of the Total Environment* **348**, 151–163.

Racioppi, F., Daskaleros, P.A., Besbelli, N., Borges, A., Deraemaeker, C., Magalini, S.I., Martinez Arrieta, R., Pulce, C., Ruggerone, M.L. and Vlachos, P. (1994) Household bleaches based on sodium hypochlorite: a review of acute toxicology and poison control center experience. *Food Chemistry and Toxicology* **32**(9), 845–861.

RHS (2004) RHS says Dig your way to health in national allotments week. http://www.rhs.org.uk/news/pressreleases/allotmentweek060804.asp (accessed 16/01/06).

Rose, J.B., Sun, G-S., Gerba, C.P. and Sinclair, N.A. (1991) Microbial quality and persistence of enteric pathogens in graywater from various household sources. *Water Research* **25**(1), 37–42.

Rusin, P., Enriquez, C., Johnson, D. and Gerba, C. (2000) Environmentally transmitted pathogens. In *Environmental Microbiology* (eds R. Maier, I. Pepper and C. Gerba), Academic Press, San Diego, CA.

Shuval, H., Lampert, Y. and Fattal, B. (1997) Development of a risk assessment approach for evaluating wastewater reuse standards for agriculture. *Water Science and Technology* **35**(11–12), 15–20.

Sibert, J.R., Lyons, R.A., Smith, B.A., Cornall, P., Sumner, V., Craven, M.A. and Kemp, A.M. (2002) Preventing deaths by drowning in the United Kingdom: have we made progress in 10 years? Population based incidence study. *British Medical Journal* **324**, 1070–1071.

Teunis, P.F.M., Heijden, van der O.G., Giessen, van der J.W.B. and Havelaar, A.H. (1996) The dose-response relation in human volunteers for gastro-intestinal pathogens. RIVM Report 284550002, Bilthoven, Netherlands.

Teunis, P.F.M., Nagelkerke, N.J.D. and Haas, C.N. (1999) Dose response models for infectious gastroenteritis. *Risk Analysis* **19**(6), 1251–1260.

Thompson, D.C. and Rivara, F.P. (2000) Pool fencing for preventing drowning in children. *Cochrane Database of Systematic Reviews* **2**, CD001047.

Tulve, N.S., Suggs, J.C., McCurdy, T., Cohen Hubal, E.A. and Moya, J. (2002) Frequency of mouthing behaviour in young children. *Journal of Exposure Analysis and Environmental Epidemiology* **12**, 259–264.

VGDHS (1999) Victoria Burden of Disease Study: Morbidity. Victorian Government Department of Human Services Melbourne, Victoria, Australia.

WHO (2002) Guidelines for drinking-water quality. Second Edition. Addendum: Microbiological agents in drinking water. World Health Organization, Geneva, Switzerland.

WHO (2004) Guidelines for Drinking-water Quality. Third Edition. Volume 1. Recommendations. World Health Organization, Geneva.

WHO (2006a) WHO Guidelines for the Safe Use of Wastewater, Excreta and Greywater. Volume IV: Excreta and Greywater in Agriculture. World Health Organization, Geneva.

WHO (2006b) Guidelines for Safe recreational Water Environments. Volume 2. Swimming Pools and Similar Environments. World Health Organization, Geneva, Switzerland.

WRAS (1999) Reclaimed water systems: Information about installing, modifying or maintaining reclaimed water systems. Water Regulations Advisory Scheme. No 9-02-04.

Yoder, J.S., Blackburn, B.G., Craun, G.F., Hill, V., Levy, D.A., Chen, N., Lee, S.H., Calderon, R.H. and Beach, M.J. (2004) Surveillance of waterborne-disease outbreaks associated with recreational water – United States, 2001–2002. *Morbidity and Mortality Weekly Report Surveillance Summaries* **53**, 1–22.

APPENDIX 5.1 SUMMARY OF INFECTION PARAMETERS

Table A5.1 shows the severity weights, durations and incidence for infection with *Salmonella* spp. and *Cryptosporidium* spp. (Mead *et al.* 1999; VGDHS 1999; Havelaar *et al.* 2000; Ruisin *et al.* 2000; CDC 2007).

Table A5.1. Severity weights, duration and incidence

Infection	Severity	Duration (days)	Incidence	Age at death
Uncomplicated salmonellosis	0.093	7	50%	
Complicated salmonellosis	0.42	10	1.1% of illness	
Fatal salmonellosis	1		0.04% of illness	78
Cryptosporidiosis (normal population)	0.054	6	71%	
Cryptosporidiosis (immunocompromised)	0.13	47	100%	
Fatal cryptosporidiosis	1		0.02% of illness	78

Table A5.2 outlines the @Risk input values for each of the assessments outlined in section 5.2.2.5.

Table A5.2. @Risk input values

Input	Distribution	Mean	Range
Salmonella shedding concentration (/g)	Log normal	10^6	10^4–10^{10}
Cryptosporidium shedding concentration (/g)	Log normal		10^6–10^7
Faecal contamination (g)	Normal	5.4	
Greywater volume (l)	Normal	105	
Volume ingested – car washing (ml)	Normal	0.5	0–1
Volume ingested – garden watering (ml)	Normal	0.5	0–1
Volume ingested – play (ml)	Normal	10	5–15
Volume ingested – external tap (ml)	Normal	125	50–250
% of households washing cars	Normal	15	10–20
% of households watering gardens	Normal	50	40–60
% of children drinking external tap water	Normal	2.5	0–5
% of households growing their own vegetables	Normal	2.5	0–5
No of children playing	Discrete		0–3
Households with salmonellosis over the summer	Discrete		3–7
Duration of *Salmonella* contamination (days)	Discrete		3–5
Households with cryptosporidiosis over the summer	Discrete		1–3
Duration of *Cryptosporidium* contamination (days)	Discrete		5–13
Log reduction by treatment	Discrete		3–6

REFERENCES

CDC (2007) Salmonellosis. http://www.cdc.gov/ncidod/dbmd/diseaseinfo/salmonellosis_g.htm (accessed 20/04/07).

Havelaar, A.H., de Hollander, A.E.M., Teunis, P.R.M., Evers, E.G., van Kranen, H.J., Versteegh, F.M., van Koten, J.E.M. and Slob, W. (2000) Balancing the risks and benefits of drinking-water disinfection: disability-adjusted life years on the scale. *Environmental Health Perspectives* **108**, 315–321.

Mead, P.S., Slutsker, L., Dietz, V., McCaig, L.F., Bresee, J.S., Shapiro, C., Griffen, P.M. and Tauxe, R.V. (1999) Food-related illness and death in the United States. *Emerging Infectious Disease* **5**(5), 607–625.

Rusin, P., Enriquez, C., Johnson, D. and Gerba, C. (2000) Environmentally transmitted pathogens. In *Environmental Microbiology* (eds R. Maier, I. Pepper and C. Gerba), Academic Press, San Diego, CA.

VGDHS (1999) Victoria Burden of Disease Study: Morbidity. Victorian Government Department of Human Services Melbourne, Victoria, Australia.

6

Flooding and health – an evaluation of the health impacts of urban pluvial flooding in the UK

Lorna Fewtrell, David Kay and Richard Ashley

6.1 INTRODUCTION

As a result of the Easter floods of 1998 in the United Kingdom and the widespread flooding occurring in both recent winter and summer months in 2007, flooding has become very topical in the UK and it has moved up on both political and research agendas. Much of this interest may be due to the perception that climate change is leading to an increase in the occurrence and magnitude of extreme flood events (Robson 2002) and that the problem is likely to recur (Few *et al.* 2004).

This chapter presents a literature review of the health impacts of flooding, and explores, through a case study, how selected impacts could be quantified for individual flood events.

The health impacts of flooding do not lend themselves to controlled prospective epidemiological studies, where those that are flooded are randomly allocated (by the scientists) and matched to a control group, who do not experience flooding. Thus, much of the literature is based on opportunist retrospective studies of flooding (sometimes conducted a considerable time after the event), which may be based on little more than anecdotal information. This results in a lot of uncertainty in the, largely qualitative, research literature. Even where quantitative data are implied, e.g. 50 people sustaining injuries as a result of a flood event, it is generally not clear what total population was affected by the event, making it impossible to quantify the resultant excess risk.

6.1.1 Health effects

Research into the health effects associated with flooding and the number of health reviews conducted (e.g. Ohl and Tapsell 2000; Hajat *et al.* 2003; Tapsell and Tunstall 2003) seems to have increased relatively recently, perhaps driven by the increase in flooding seen during the twentieth century (Milly *et al.* 2002) and the forecast from climate change modelling that this trend will continue (Evans *et al.* 2004). In the UK, research is being sponsored by a number of agencies, including The Department for Environment Food and Rural Affairs (DEFRA) and the Environment Agency (EA) as part of their joint Flood and Coastal Defence research and development programme and the Engineering and Physical Sciences Research Council (EPSRC).

No geographical restriction was placed on the literature review and relevant English language papers (or abstracts) from any country were examined. The health outcomes resulting from floods in developing countries tend to be more numerous and may be more serious (with a greater number of people drowning and suffering from outbreaks of illness), but have not been detailed here unless they are thought to be relevant to conditions that exist in the UK.

The health effects of flooding are generally split into those associated with the immediate event (with drowning being the most obvious) and those arising after the flood has resolved (i.e. post-onset, which may be related to exposure to floodwaters, the clear up process or stress and anxiety). Health impacts can also be considered to be direct or indirect, with indirect impacts perhaps resulting from damage to infrastructure, food supply and so on. Indirect impacts are not considered here, except in the consideration of flooded properties with private water supplies.

Health effects have been categorised into a number of groups, namely:

- mortality and injuries;
- infection; and
- mental health effects.

A number of studies have examined flood events for multiple health outcomes, with effects falling into more than one of these categories. These studies have been referred to in each appropriate section.

6.2 MORTALITY AND INJURIES

This section examines mortality and injuries as an immediate direct consequence of flooding, as opposed to, say, flooding acting as a stressor on a vulnerable person and a subsequent death, some months afterwards, from a heart attack (hastened mortality). Injuries may occur during the pre-onset and onset phase of flooding (when people are trying to escape or remove valued possessions from the approaching floodwaters) or during the post-onset phase (when people return to their property and start the clean up process (Few *et al.* 2004).

6.2.1 Mortality

Examination of the EM-DAT database (on the occurrence and immediate effects of all reported mass disasters), which was established in 1988, reveals that the UK has suffered from 17 flooding events between 1900 and 2005 that warrant an entry[1]. A total of 51 people were reported killed and eight injured in these events, with over 4000 people affected by the events (EM-DAT 2005). This, however, does not seem to be comprehensive as it does not include the 1953 storm surge affecting coastal areas of several North Sea countries, including the UK. Although reported death tolls vary according to which source is examined, it would seem that over 300 were killed on land, during the event, with a further 200+ killed at sea (Kelman 2003a). In addition, 20 people were killed in eastern England in floods on January 11th, 1978, and five people were killed during the

[1] For entry at least one of the following criteria must be satisfied: 10 or more people reported killed, 100 people reported affected, a call for international assistance or declaration of a state of emergency. It should be noted, however, that not all of the UK flood examples in this database seem to fall into one of these categories.

April flooding in 1998 (Kelman 2003b). Analysis of events surrounding the autumn 2000 floods suggests that four drownings may have been related to the flood and the fact that there weren't many directly attributable to the event seems to be more the result of luck than judgement (Kelman 2003b), as many people are, perhaps not surprisingly, ignorant of appropriate behaviour during flooding.

Duclos and Isaccson (1987) report on 24 deaths related to flooding events in the USA, only nine of which had drowning as the primary cause. Another nine were due to heart attacks while people were involved with unusual flood-related stress or activities (including moving furniture and clean up processes). The other two deaths relevant to the flood event were asphyxiation while using a gas generator to pump water out of a basement and electrocution while connecting a pump in a flooded basement. In flash floods affecting Georgia, 28 deaths were classified as flood-related (CDC 1994a), 27 of these deaths were due to drowning (cause of death was unknown in one case), with 20 being motor-vehicle related (i.e. victims drove into low-lying areas, across washed out bridges or off the road into deep water). Enhanced surveillance of emergency departments was conducted in North Carolina, USA, following Hurricane Floyd (CDC 2000a). In total, 52 deaths were attributed to the event with drowning being the most common cause (Table 6.1). Five of the deaths were in rescue workers.

Table 6.1. Deaths related to Hurricane Floyd, by cause of death (CDC 2000a)

Cause of death	Number	%
Drowning	36	69
In motor vehicle	24	
In boat	7	
As pedestrian	4	
In house	1	
Motor-vehicle crash (excluding drowning)	7	13
Myocardial infarction (heart attack)	4	8
Fire (burns and trauma from escape attempts)	2	4
Hypothermia	1	2
Electrocution	1	2
Fall	1	2

More recently (August 2005), parts of the USA (Mississippi, Louisiana, Alabama and Florida) were affected by Hurricane Katrina. Preliminary mortality reports suggest that approximately 1000 hurricane-related deaths occurred in Louisiana, 200 in Mississippi and 20 in other States (CDC 2006a). The majority of deaths resulted from storm surges along the Mississippi and Louisiana coastlines and flooding in the New Orleans area.

In France, Duclos *et al.* (1991) reported details of a flash flood that occurred in the region of Nîmes. Despite extensive flooding and a great deal of damage only nine deaths were reported (all due to drowning) and three severe injuries. The low impact in terms of mortality and serious injuries was attributed to the time of flooding (most people were still at home).

During 1997, Poland experienced massive flooding, with water covering approximately 10% of the landmass. It is thought than 3 million people were affected and it has been estimated that due to the rapidity of the flooding some 55 people were killed (Riddez and Dellgar 2005).

6.2.2 Injuries

Injuries may occur during flooding (e.g. from impact with flood-related debris or being knocked over by the floodwater) and also from the subsequent clean up process. During surveillance of emergency departments and enhanced surveillance (including contact with emergency shelters) following the Midwest flood in Missouri, USA, in 1993, 524 flood-related conditions were reported, 250 (47.7%) of which were injuries, with the most common being sprains, strains and lacerations (CDC 1993). It is estimated that 60,000 people were displaced during these floods – giving a reported injury rate of 0.4%. Following Hurricane Floyd in the USA, surveillance of emergency departments recorded a variety of flood-related visits, many of which were related to orthopaedic and soft tissue injury (28%). Comparison of injury data following the hurricane with pre-hurricane emergency department visits found an increase in dog bites, insect bites and violence (CDC 2000a). A survey of police officers and fire fighters in New Orleans, following Hurricane Katrina, found that lacerations and sprains were the most common injuries (CDC 2006b). A total of 22% of those surveyed reported lacerations (311/1437), 17% reported sprains or strains and 10% reported animal bites or stings. Sullivent *et al.* (2006) also examined non-fatal injuries in residents and relief workers over a five week period shortly after the hurricane struck. Over 7500 injuries were recorded with falls and cut/stab/pierce injuries being the most common. Injuries were most frequently associated with the clean-up process.

In France, Duclos *et al.* (1991) reported three serious injuries as a result of flooding (arm and face burns, a broken leg, broken arms) and found that 6% of occupants in households surveyed following the flooding in Nîmes reported mild injuries (such as bruises, cuts and sprains).

Daley *et al.* (2001) reported an 'outbreak' of carbon monoxide poisoning that resulted from the use of petrol-powered pressure washers for the cleaning of flooded basements. There were 18 separate incidents, involving 33 cases. Thirty of the cases reported that the basement was ventilated while the pressure washer was in use. The most common symptoms were dizziness, headache, nausea and fatigue. With the exception of a patient who lost consciousness, all the cases were released from hospital after receiving oxygen.

6.2.3 Mortality and injury quantification

In the 'Flood risks to people' research project (DEFRA/EA 2004, 2006) a flood hazard formula for assessing risks to people has been devised based upon deaths and 'serious harm', which has terms for depth of water, water velocity and a debris factor, as well as area vulnerability and people vulnerability. This was developed with flooding from rivers, estuaries and the sea in mind. Relatively minor injuries will be quantified by assuming that 6% of occupants of flooded properties are affected by cuts, bruises and sprains (based on the figure from Duclos *et al.* 1991).

6.3 INFECTION

A number of infections can result from contact with floodwater. These can be caused by a variety of different agents (bacterial, viral, protozoal and fungal), they can occur via a range of exposure routes (e.g. ingestion, inhalation, wound infection, insect transmission), with the severity of infection ranging from sub-clinical or mild to fatal. In developed countries, fatal infections are probably rare, although a case was reported from the USA (Spice, 2004) where a 57 year old man died from septicaemia (blood poisoning). The infection was thought to have been the result of contact between floodwaters and an open wound on his leg; the victim was also diabetic, which would have made him more susceptible to infection.

The following sections are split, where possible, by the infective agent (e.g. bacterial) or the transmission route (e.g. vector-borne infections). In some cases, such as gastrointestinal illness and respiratory infection, where no causative agent was determined, these have been included as separate sections.

6.3.1 Bacterial

6.3.1.1 Legionella

In the USA, the owner, a waitress and a regular customer of a bar became ill with Legionnaires' disease (Kool *et al.* 1998). Investigations revealed that the only common exposure for these cases was the bar, which had been flooded nine days before the first reported illness. *Legionella pneumophila*, of the same serotype isolated from the cases, was found from the sump in the crawlspace under the bar and from a hose running from the sump to the street. Following the flooding, water was pumped from beneath the bar to the street over a period of three days by an old electrical sump pump. The pump generated a considerable amount of heat while running and it is thought that this may have provided temperatures favourable to the growth of *Legionella*. This, in conjunction with an aerosol generated from a jet of water coming from a small hole in the side of the pump, was probably responsible for the outbreak. These are relatively unusual circumstances, but do illustrate the potential for the transmission of *Legionella* following flooding.

6.3.1.2 Leptospirosis

Flooding has been reported as a risk factor related to leptospirosis infection, principally in people with skin cuts or abrasions coming into contact with floodwater (e.g. in Mexico – Leal-Castellanos *et al.* 2003). In the USA, outbreaks of illness have also been reported following flooding (Fuortes and Nettleman 1994; WHO 2005). Here, it was thought that environmental changes increased the local rodent populations which facilitated transmission. In the Czech Republic, flooding in 1997 and 2002 was followed by a three fold increase in reported cases of leptospirosis (Zitek and Benes 2005). The majority of cases were from areas that were inundated and 50% could be directly associated with the floods (i.e. there was exposure to residual water or flood mud). Although similar research has not been identified in the UK, recreational contact with surface water is known to be a risk factor (Fewtrell 1991) and, thus, precautions against leptospirosis (such as covering cuts) when in contact with floodwater would be prudent.

6.3.1.3 Quantification of bacterial infection

Legionella

Only a single instance of legionellosis as a result of flooding has been identified from the literature and this resulted from the culmination of a series of specific circumstances (Kool *et al.* 1998). *Legionella* spp., although ubiquitous in the water environment, require warm (20–50 °C) temperatures in which to grow. These are unlikely to be realised in environmental floodwaters under current UK climatic conditions. Additionally, aerosol generation is required for transmission, therefore, it is assumed that legionellosis is unlikely to result from pluvial or coastal flooding in the UK, and so this has not been quantified.

Leptospirosis

Although flooding may spread the organism (because, as water saturates the soil, the leptospires pass directly into surface water) leptospirosis is primarily a disease of tropical and subtropical regions. It is uncommon in the UK, with only 37 cases reported in England and Wales during 2004 (CDR 2005a), representing an incidence rate of 0.07/100,000 population. Thus, under current climate conditions, it is not considered that leptospirosis will result from flooding.

6.3.2 Fungal

Fungi (which can encompass mushrooms, yeasts, toadstools, rusts, moulds and mildews) are common in the environment (Kibble *et al.* 2003). A number of moulds, which may be found in indoor environments where there are high levels of moisture (such as in water-damaged properties), may be considered toxic (producing toxic mycotoxins). Exposure is usually via inhalation of airborne spores or direct contact. Exposure via ingestion of contaminated food is also possible but is unlikely to be related to flooded domestic properties.

Much of the concern relating to fungus and flooding is linked to a series of reports, which associated the presence of *Stachybotrys chartarum* (previously known as *S. atra*) in water-damaged homes with pulmonary haemorrhage (bleeding of the lungs) in infants (CDC 1994a and b; Dearborn *et al.* 1997). This evidence was seemingly strengthened by Elidemir *et al.* (1999), who reported the first isolation of the toxigenic (producing a toxin or toxic effect) fungus, *S. atra*, from a child with pulmonary haemorrhage and environmental samples from his water-damaged home. The patient had moved to a severely flood-damaged home shortly before his symptoms started. Environmental investigation revealed dense areas of black-coloured mould, which culture revealed to include *S. atra*. Cleaning of the house according to guidelines produced by the Centers for Disease Control and Prevention (CDC) and relocation of the patient to a non flood-damaged home for two months led to the resolution of his symptoms.

In 2000, however, CDC (2000b) noted that there had been shortcomings in both the implementation and the reporting of the investigations previously outlining the association between pulmonary haemorrhage and *S. chartarum* (CDC 1994a and b; Dearborn *et al.* 1997). Although associations have been noted there is currently limited evidence for causation (Kuhn & Ghannoum 2003; Hardin *et al.* 2003) and so this area has not been considered further, other than in general terms under respiratory disease.

6.3.3 Respiratory disease

Dales *et al.* (1991a and b) examined the effects on health in adults and children (between the ages of 5 and 8) exposed to home dampness and moulds (defined as damp spots, visible mould or mildew, water damage and flooding). In adults, lower respiratory symptoms (any cough, phlegm, wheeze or wheeze with dyspnoea – laboured or difficult breathing) were reported more frequently in those living in homes with dampness or mould (Dales *et al.* 1991a). The relationship remained after adjusting for a number of socio-demographic and exposure variables (adjusted odds ratio [OR] 1.62; 95% confidence interval [CI] 1.48–1.78). In children, the prevalence of all respiratory symptoms was higher in homes with reported dampness or moulds. The adjusted odds ratios ranged from 1.32 (95% CI 1.06–1.39) for bronchitis to 1.89 (95% CI 1.58–2.26) for cough (Dales *et al.* 1991b). Although these studies were conducted as questionnaire surveys and no fungal spore measurements were taken, previous studies have suggested that respondents' perceptions of home dampness correlate with measured spore concentrations (von Wageningen *et al.* 1987) and that where a 'water problem' existed for more than three days this was associated with an increase in measured spore levels (Gallup *et al.* 1987). It is not known whether this North American research is likely to be applicable to the UK situation, and its relevance to flood events is uncertain.

Research conducted in the UK by the Flood Hazard Research Centre (FHRC), has suggested that respondents at focus group meetings following flooding in the northeast of England (June 2000) were likely to report asthma, chest infections, coughs, colds and flu believed to be due to the flooding (Tapsell *et al.* 2002), although the specific symptoms reported vary by the area studied. In a DEFRA/EA study (DEFRA/EA 2003 – reported more fully in section 6.4.2), respondents were asked to report on physical health effects following flooding. Respiratory symptoms (namely colds, cough, sore throats and flu) were reported by 20% of the respondents. Reacher *et al.* (2004), in a study of the health impacts experienced as a result of flooding in Lewes (reported more fully in sections 6.3.5 and 6.4.1.2), found a significant association between flooding and self-reported worsening of asthma in adults (relative risk [RR] 3.1; 95% CI 1.2–4.4) but not for respiratory illness in children or adults.

6.3.3.1 Quantification of respiratory illness

The figures above, on the worsening of asthma in adults, have been used in an attempt to quantify the impact of flooding on respiratory health. In order to calculate any worsening, it is necessary to know the number of people with asthma. Recent figures suggest that 13.8% of people in the UK suffer from asthma and that approximately 9% of UK residents are receiving long-term treatment for asthma (European Opinion Research Group 2003; Hoare *et al.* 2003), with the rate of treatment being higher in children than adults. Thus, the relative risk determined by Reacher *et al.* (2004) has been applied to 7.6 % of the adult population affected by flooding[2].

6.3.4 Gastrointestinal infection

Aavitsland *et al.* (1996) examined illness levels in eastern Norway, following a river flood during 1995, when over 7000 people were forced to leave their homes. They observed no increase in gastrointestinal illness, although they did note that advice and public information on the safe management of floodwater had been communicated to the affected population, and measures were also taken to protect water supplies. Ogden *et al.* (2001) also found no apparent increase in waterborne or infectious disease following flooding in Louisiana, based on reports from emergency departments.

A study conducted in the northeast of England found that many people who had been flooded reported diarrhoea and upset stomach symptoms following the flooding (Tapsell *et al.* 2002), although the study was not designed to determine whether these effects were statistically significant.

Following tropical storm 'Allison' in Houston, Texas (USA) in June 2001, the Department of Health and Human Services conducted a rapid needs assessment in the areas most affected by flooding (CDC 2002). The survey took place one week after the heaviest rainfall. Questionnaires were completed by people from 420 households, 137 had floodwater in the home during the event. Almost 13% of the surveyed households reported at least one person with illness that occurred after the onset of the flood and people living in flooded homes were significantly more likely to report illness than those from non-flooded homes (OR 4.7; 95% CI 1.8–12.0). Gastrointestinal illness in people from flooded homes was especially elevated (OR 6.2; 95% CI 1.4–28.0).

In a retrospective study of self-reported health impacts arising from riverine flooding in Lewes, East Sussex, Reacher *et al.* 2004 did not find that gastroenteritis was a statistically significant health outcome, when comparing

---

[2] Approximately 5.1 million people in the UK receive treatment for asthma, of which 1.4 million are children. Using figures from the National statistics website for a population of 48.7 million adults and 10.9 million children (aged less than 15) suggests that 7.6% of adults are receiving long-term treatment for asthma.

flooded with non-flooded properties (RR 1.7; 95% CI 0.9 – 3.0), although there was a relationship with the depth of flooding (p for trend = 0.04).

The flooding component of a study by Wade *et al.* (2004) was an opportunist addition to an ongoing project studying the incidence of gastroenteritis related to drinking water. Flooding along parts of the Mississippi River occurred over a period of six weeks in April and May 2001; during some of this time the floodwater would have contained untreated sewage and wastewater. After the flood had subsided participants in the study were sent a questionnaire asking them about the extent to which they had come into contact with floodwater or flood-contaminated items. Health data continued to be collected on a routine basis. The rate of diarrhoea resulting in a doctor's visit; days of missed work or school due to gastrointestinal symptoms; and days of vomiting were not elevated following flooding. There was, however, an overall elevation in the reporting of highly credible gastrointestinal (HCGI) symptoms[3] during the flood period in comparison with other periods (incidence rate ratio [IRR] 1.29; 95% CI 1.06–1.58). This was particularly marked in those people who reported flooding in their house or yard (IRR 2.36; 95% CI 1.37–4.07), especially in children aged 12 or less (IRR 2.42; 95% CI 1.22–4.82). As noted by the authors, direct contact with floodwater did not entirely explain the increase in incidence of gastrointestinal symptoms observed during the flood period, as the incidence was still elevated in people who had no reported floodwater contact.

Where properties have private drinking water supplies, flooding may result in contamination of the source water with a variety of microorganisms including gastrointestinal pathogens, such as *Cryptosporidium* spp. (Petrie *et al.* 1994; Duke *et al.* 1996; Fewtrell *et al.* 1998; DWI 2003). Increased turbidity in the source water as a result of flooding may overwhelm treatment systems and, in such cases, where properties are still occupied and water is not boiled or chemically treated an increase in gastrointestinal illness is likely.

6.3.4.1 Gastrointestinal illness quantification

Given the range of figures reported in the literature, from no apparent increase in gastroenteritis, up to an OR of 6.2 (95% CI 1.4–28), it is difficult to know what to choose for quantification. The study by Wade *et al.* (2004), however, provides a reasonable compromise with an IRR of 2.36 (95% CI 1.37–4.07) and this has been used in the case study health impact evaluation. An elevation in serious gastrointestinal problems, however were not reported, so the duration and

[3] HCGI symptoms included any of the following: liquid diarrhoea, soft diarrhoea and cramps, nausea and cramps, vomiting.

severity of cases has been assumed to be short and mild, respectively. Estimates of the typical number of episodes of gastrointestinal illness per person per year in developed countries vary, with a range of 0.03–2.6 (Payment and Hunter 2001). Hence, for simplicity, a figure of 1 episode/person/year has been assumed as a general incidence rate. Effects on private water supplies are site and event specific and have not been considered in the health impact evaluation as the case study areas (see section 6.5) are not served by private supplies.

6.3.5 Earache and skin rash

Reacher and colleagues performed a historical cohort study using telephone interviews in residents from 103 flooded and 104 non flooded properties in the same postcode area in Lewes, Southern England (Reacher *et al.* 2004). The telephone interviews were conducted nine months after the flooding; information was recorded for injuries, worsening of pre-existing asthma, respiratory illness, earache, gastroenteritis, skin rash and, for adults, psychological health. In subjects of all ages, earache was significantly associated with having been flooded (RR 2.2; 95% CI 1.1–4.1), there was a weaker, non significant association between flooding and skin rash.

6.3.5.1 Quantification

Earache can be quantified using the relative risk determined by Reacher *et al.* (2004). No details are given on duration or severity so these have been assumed to be short (three days) and mild. The incidence of earache is unknown, but the incidence of acute otitis media (inflammation of the middle ear) in the Netherlands in general practice is about 20 per 1000 patients (Appelman *et al.* 1999). Given that fewer people would be expected to visit their general practitioner for less serious and short-lived earache a general incidence of 50/1000 has been assumed to apply to the UK population (these figures do not account for number of GP visits but are assumed to be a general population annual incidence). As skin rash was not found to be significantly elevated this has not been quantified.

6.3.6 Vector-borne illness

A number of insect vectors, most notably mosquitoes, can transmit illnesses including malaria and a number of viral infections (such as West Nile virus). As mosquitoes require pools of still water for breeding, flooding may provide an increase in possible breeding sites. Several studies have examined the effect of flooding on the number of mosquitoes in the flooded area and also transmission of illness. In Iowa, USA, unusually high levels of summer rainfall led to an increase in the mosquito population in the area. Gilliland *et al.* (1995) collected mosquitoes and examined pooled samples for the presence of arbovirus (of which some types can cause encephalitis [inflammation of the brain]). Although a number of arbovirus isolates were found from the mosquitoes, sentinel chicken populations (established to examine the possibility of transmission of arbovirus) and human surveillance suggested that there was no increase in virus transmission. A similar surveillance operation conducted in southern regions of the Midwest flood area in 1993 found arbovirus (western equine encephalitis) in only a single pooled sample (CDC 1994c).

In July 1997, flooding which occurred after heavy rains in the Moravia area of the Czech Republic led to a dramatic increase in mosquito populations (Hubalek *et al.* 1999). Tahyna virus and West Nile virus (WNV) were isolated from a sample of mosquitoes and there was some indication of human infection with these agents, with two children having an illness compatible with West Nile fever. Han *et al.* 1999 reported that living in apartment blocks with flooded basements in Bucharest, Romania, was a risk factor for WNV infection. The basements were, however, flooded with sewage leaking from poorly maintained plumbing, which created a high-organic habitat favourable to the mosquito vector.

6.3.6.1 Vector-borne illness quantification

Malaria

Malaria is not currently indigenous to the UK, and rare cases of 'airport malaria' are caused by infected mosquitoes arriving from malaria endemic countries on planes and surviving for long enough to bite someone (Guillet *et al.* 1998; Mouchet 2000). Thus, under current UK conditions, no cases of malaria are expected to result from indigenous flooding.

Arbovirus

The most commonly discussed arbovirus is WNV, which became the focus of public and scientific attention following outbreaks of human illness (and fatalities) in the previously non-endemic USA (Fewtrell 2002). Birds are the normal hosts and act as a reservoir of infection, with humans and domestic animals being incidental or 'dead-end' hosts. The mosquito *Culex pipiens* is thought to be the most likely for the establishment of WNV in the UK (Higgs *et al.* 2004). Although studies have found the presence of WNV-neutralising antibodies in UK-resident birds, suggesting that the virus may have been introduced by migratory birds (Buckley *et al.* 2003), no virus has been isolated from either birds or mosquitoes in the UK (Medlock *et al.* 2005). Enhanced surveillance for human infection acquired in the UK has been operating during the summer months since 2001 (CDR 2005b). To date, no human cases have been reported in the UK.

6.4 MENTAL HEALTH EFFECTS

There has been increasing interest in the possible mental health effects of flooding events. Mental health impacts vary in severity and include a wide range of symptoms, including possible suicide attempts, post-traumatic stress disorder (PTSD – characterised by intrusive memories, avoidance behaviour, sleep disturbances, irritability, anger and excessive vigilance), depression and substance abuse. The studies identified from the literature have been divided according to whether they examined a single flooding event, or multiple events.

6.4.1 Single event studies

A number of retrospective studies examining single events have been identified from the literature. The earlier studies typically examined a flooded population and sought information on symptoms or GP visits in the 12 months prior to, and 12 months after, the flood event, either through self-reported symptoms or GP records (Bennet 1970; Abrahams *et al.* 1976). Reacher *et al.* (2004), however, in the most recent UK-based quantitative study, simply compared a flooded group and a control group on the basis of their experiences after the flood. In France, Duclos *et al.* (1991) examined the immediate health impact of the flash flood in Nîmes, without using a control group. Studies often included a variety of physical illnesses (which have been outlined in earlier sections) as well as mental health effects.

6.4.1.1 Post-traumatic stress disorder

Researchers in Bangladesh were able to undertake a 'natural' experiment examining the effect of flooding on child behaviour (Durkin *et al.* 1993). A population-based study, of behavioural problems in over 2500 children aged between 2 and 9 years old, had been conducted six months before a major flood event. Five months after the flood (i.e. 11 months after the original study), 162 children (49 with and 113 without behavioural problems) were re-evaluated. After the flood, 16 children were newly reported to be very aggressive and incontinence increased from 16.8% to 40.4%. The increases in both aggression and incontinence were seen among both children with and without disabilities. Aggressive behaviour increased at nearly every age. The authors suggest that these results help to define what may be considered symptoms of post-traumatic stress distress in childhood.

Verger *et al.* (1999, 2000) examined the risk factors for post-traumatic stress disorder symptoms five years after a large flood in the Vaucluse area of France. The flood, which affected over 60 municipalities, resulted in 38 deaths. In a pilot study, the researchers conducted face-to-face interviews with residents from two different areas affected by the flooding and found that the PTSD scale (based on a standardized questionnaire) was an effective tool (Verger *et al.* 1999). In the main study, a single area (Bedarrides) was targeted and 500 people interviewed. Significantly higher PTSD scores were observed for females, people over the age of 35, with a low monthly income and those who had a history of psychological disorders or life-threatening traumatic events. They also found a significant exposure-response relationship with the level of exposure to the flood and PTSD score (Verger *et al.* 2000).

In Canada, Auger *et al.* (2000) conducted a telephone survey of flooded and non flooded areas in Saguenay four months after a flood event. A total of 141 adults were interviewed; 62 from the flood area and 79 from an adjacent non-flooded area. PTSD was found to be significantly elevated in the flooded group (OR 6.08; 95% CI 1.63–22.64), with prevalence in the flooded group being almost 20% compared to 3.8% in the control group. Emotional distress was also significantly higher in the flooded group (OR 2.42; 95% CI 1.04–5.61).

McMillen *at al.* (2002) studied people who had been affected by the Midwest floods in 1993, which resulted in 50 deaths, damage to 50,000 homes and the evacuation of 54,000 people. Interviews were conducted with 160 people (from those thought to be worst affected by the flooding), 40 of whom reported flood-related injuries. Some 16 of the respondents saw others injured or killed in the flood, 25 had a family member injured or killed in the flood and 118 people reported doing or seeing something disgusting during the flood. McMillan *et al.* (2002) found that 60 people (38%) met criteria for a post-flood psychiatric disorder and that 35 people (22%) met criteria for PTSD related to the flood, although it should be noted that the response rate to the invitation to participate in the study was only 30%.

In hurricane Katrina evacuees, who were willing to talk about their experiences, 41% said they were afraid they would die and 16% saw someone close to them injured or die. Using an Impact of Stress Experiences scale, Coker *et al.* (2006) found that the majority of the evacuees scored as experiencing moderate (38.6%) to severe (23.9%) PTSD.

In Poland, Bokszczanin (2002) found that exposure to the 1997 flood disaster was a strong predictor of symptoms of PTSD, depression and feelings of loneliness in children and adolescents. The study, which took place 20 months after the flood, collected questionnaire response data from children and adolescents between the ages of 11 and 20. The occurrence of more intense symptoms was seen in those who had suffered greater threats during the flood event (fear of death, evacuation, being trapped by the flood and so on).

6.4.1.2 General symptoms

Clemens *et al.* (1999) examined the effect of flooding on the risk of domestic violence in the USA, following a flood in North Dakota. A retrospective cross-sectional survey was conducted on 140 adults, which showed that domestic violence was significantly greater among respondents after the flood. It was also shown that flooding led to increased levels of anxiety, depression and hostility, although these symptoms did not automatically lead to domestic violence.

In the USA, Ginexi *et al.* (2000) studied Iowa residents before and after the 1993 Midwest floods. An Iowa Health Poll (IHP) conducted in 1992 (with 2379 respondents), a year before the flooding event, was used to provide pre-disaster data. Attempts were made to contact those who had participated in IHP, two – three months after the flooding, to provide post-disaster data (Iowa Flood Study – IFS). The researchers were successful in contacting and re-interviewing over 70% of the original respondents. Although the whole of Iowa had been declared a Federal disaster area as a result of the flood, the actual level of flooding experienced by people varied markedly. Over 50% of the respondents to the IFS reported some flood impact but only 2% of the respondents had vacated their homes. Despite this, the study was able to demonstrate a small but true rise in depressive symptoms in those affected by the flood.

Reacher and colleagues performed a historical cohort study using telephone interviews to establish physical illness and psychological distress in residents from 103 flooded and 104 non flooded properties in the same postcode area in Lewes, Southern England (Reacher *et al.* 2004). Previous studies (Bennet 1970; Abrahams *et al.* 1976) had suggested that psychological distress may be associated with flooding, so this was specifically examined in adults using the GHQ-12 (a 12-item General Health Questionnaire, which is a standardised tool used to assess psychological symptoms). At the peak of the flooding, some areas of the town were under 3.6 m of water. No deaths or trauma were reported as a result of the flood and routine surveillance of infectious disease did not detect an increase in the aftermath of the flood. The telephone interviews were conducted nine months after the flooding; information was recorded for injuries, other physical symptoms (see earlier sections) and, for adults, psychological health. Psychological distress in adults (pre-defined as a score of ≥4 in response to the GHQ-12) was significantly raised in the flooded group (RR 4.1; 95% CI 2.6–6.4). Psychological distress also exhibited a significant trend by flood depth. There were marked decreases in the risk estimates for physical illnesses in adults after adjustment for psychological distress. The authors note that their results suggest that the risk of psychological distress from being flooded was independent of reported physical illness and sanitary disruption within flooded households, they suggest: "*Our findings support the view that psychological distress is a leading adverse health outcome associated with flooding and may explain a proportion of physical illness.*"

6.4.2 Multiple event studies

In 2003, DEFRA and the EA published a study on the appraisal of human-related intangible impacts of flooding, much of which was aimed at producing a methodology to account for the impacts of flooding on human health in order to assess the benefits of flood alleviation measures (DEFRA/EA 2003). Questionnaires were developed and a number of previously used scales designed to assess health effects. The scales considered the most appropriate to flooding and a UK population after pilot trials were the GHQ-12 and PTSS (Post-Traumatic Stress Scale). The SF-12 (Short Form) could not be used as the group were unable to obtain formal approval for its use.

In the main study, over 1500 face-to-face interviews in 30 locations across England and Wales were conducted with people who had been flooded (983) or were at risk of flooding (527). The questionnaires were used to assess both health status at the time of the interview (considered to be an indicator of the long-term effects of flooding) and health at the time when the flooding was most severe for them (an indicator of the short-term effects of flooding). The number of people who had been flooded scoring 4 (the threshold indicative of some degree of mental health problems) or greater on the GHQ-12 was high, with 64% of all respondents recording this level at the worst time of flooding (generally within three months of the event), and 25% when current health was being assessed. In comparison, only 10% of those 'at risk' of flooding scored 4 or more. Interestingly, the figure of 10% is less than that reported by the general population, for whom the proportions scoring four or more on this scale were reported to be 13% for men and 18% for women (DoH 1999). Most of the respondents (72%) reported experiencing some psychological effects as a result of the flooding, with most people reporting an increase in anxiety during rainfall events. In subjective terms, stress rated highly in terms of the severity of the effects of flooding, while health effects (presumably, other than stress) were rated among the lesser effects of flooding.

Three factors were considered in relation to the effects on health:

- socio-demographic factors;
- flood characteristics; and
- post-flood factors.

The socio-demographic factors found to be important were gender, prior health, age, rented housing, flood risk awareness and long-term illness, although the ranking of these depended upon the health measure used (PTSS, current GHQ-12 or worst time GHQ-12). The flood characteristics found to be important were contamination, depth of flooding, years since flooding and warning time. Problems with insurers and evacuation were found to be significant post-flood factors (DEFRA/EA 2003; Tunstall *et al.* 2006).

In a series of studies conducted by the FHRC on behalf of the EA in response to the Easter floods of 1998 and the summer floods of 2000, the following self-reported psychological health effects were reported by those who had been flooded (Tapsell and Tunstall 2003):

- anxiety (e.g. during heavy rainfall);
- increased stress levels;
- sleeping problems;
- depression;
- panic attacks;
- flashbacks to flood;
- difficulty concentrating on everyday tasks;
- lethargy/lack of energy;
- feelings of isolation;
- increased use of alcohol or prescription (other) drugs;
- nightmares;
- anger/tantrums;
- mood swings/bad moods;
- increased tensions in relationships (e.g. more arguing); and
- thoughts of suicide.

6.4.3 Mental health quantification

Mental health is a wide ranging and complex area, which is especially difficult to quantify. However, in order to try to capture at least some of the impact on people psychological distress in adults has been used as the health outcome, with the relative risk of 4.1, derived by Reacher *et al.* (2004), being assumed to apply to those flooded. A baseline incidence of 15.5% has been used, based on the average figure for the general population (13% for men and 18% for women) reported in DoH (1999). As research suggests that these effects are long lasting, the duration has been assumed to be one year. There is no specific severity weighting in the literature for this category, although there are weightings for a number of anxiety disorders, including mild to moderate PTSD (with a weighting of 0.130) and mild attention-deficit hyperactivity disorder (a childhood condition), with a weighting of 0.02 (VGDHS 1999). It has been assumed that a severity weight of half of that for mild PTSD is appropriate (0.065).

6.5 HEALTH IMPACT EVALUATION – BRADFORD CASE STUDIES

Two areas in Keighley (West Yorkshire, UK), Utley and Devonshire Park, were selected as case study examples, as both areas have been subjected to flooding on at least one occasion within the last five years. A socio-demographic profile for each case study location was created using the *Map and Stats* website (http://www.mapandstats.com) which allows incomplete output areas to be combined from user-defined areas in this region (providing either a certain number of houses are selected or a target population is achieved – to maintain the individual anonymity provided by output areas). In addition to local knowledge, the Bradford urban flood questionnaire was issued to households and businesses likely to have been affected by flooding in the two areas. This was designed to acquire a range of flooding data, including information on flood depth, frequency of flooding and health impacts, to create a database of flood information.

6.5.1 Utley

This area is located to the north east of Keighley town centre. The ground slopes from west to east. The area is served predominantly by a combined drainage system, but there is a complex, though poorly defined, urban drainage system, much of which has been culverted. During extreme rainfall, water from urban greenspace and farmland flows into the urban area and it is estimated from local knowledge and questionnaire data, that 20 properties are typically flooded. The depth of floodwater from the most recent event was estimated to be 0.1 m and the velocity 1.5 m/s.

The Utley case study area as a whole has a population of 2041 based on 2001 census figures, with fewer children and people under the age of 24, more people over the age of 45 and with more all-pensioner households than the Bradford District as a whole. Also, in comparison with the Bradford District, it has a greater white population, more owner occupiers, more detached and terraced houses, less unemployment, a greater proportion of retired people and the residents are more likely to have a professional or degree level qualification. The average household size is 2.24 people. Closer examination of the two residential areas (Hollins Lane and the Beechcliffe area), which have reported past flooding, revealed some notable differences, so these have been examined separately.

The Hollins Lane area has a population of 261 (it is not possible to look at a smaller area), with a high proportion of residents aged between 45 and 64. Households are far less likely to have no amenities and be without a car than the other case study areas or Bradford as a whole. People were also more likely to have better health. Over 50% of dwellings are detached and the majority of properties are owner occupied. The average household size is 2.4 people. Baseline conditions suggest that 12 houses are flooded with a return interval of ten years.

In contrast, the Beechcliffe area, with a population of 226, is much more deprived. The area has more households without amenities than Bradford as a whole, with a high proportion of terraced houses and fewer residents with professional qualifications than Bradford or Hollins Lane. The average household size is 2.4 people. Baseline conditions suggest that 8 houses are flooded with a return interval of ten years.

6.5.2 Devonshire Park

Devonshire Park is located to the north east of Keighley town centre. The ground slopes from west to east. The area is served by a combined drainage system. Many residents and property owners have protected their properties by means of upstands[4] at entrances to driveways, but some properties have been flooded to a considerable depth and sheltered accommodation has also been affected. It is estimated that typically 10 properties are flooded, with a return interval of ten years. The depth of floodwater during the most recent event was estimated to be 0.1 m and the velocity 1.5 m/s.

The Devonshire Park case study area has a population of 553. In comparison with the Bradford district the area has more children and young adults, less people over the age of 25 and less elderly (although this excludes the residential home within the area boundaries). The residential accommodation is home to

[4] i.e. small concrete ramps to reduce water entry.

about 58 residents, with an average age over 80. The area has more overcrowding, more households without amenities and a greater percentage of people with a long term limiting illness compared to both the Bradford district and the Utley area. Residents are mainly Pakistani (68%) and less likely to be owner occupiers (57.9% compared to 71.7% for Bradford District). There is more unemployment and people are more likely to have no qualifications. The average household size is 3.67 people.

6.5.3 Quantification methodology

The quantification of health impacts is based principally upon:

- '*Flood risks to people*' research project (DEFRA/EA 2004, 2006) for estimates of deaths and serious injuries; and the
- WHO methodology, adapted from the comparative risk assessment calculations of global disease burden (Prüss *et al.* 2002), which combines incidence in the general population with relative risk values derived from the literature.

Figures generated from each of these components were converted into disability adjusted life years (DALYs), by combining the estimates with figures for severity and duration (See Chapter 2).

6.5.3.1 Flood risks to people

The 'Flood risks to people' project developed a formula to estimate death and serious harm occurring as a direct result of flooding either during a flood event or up to one week after the event. The formula has a number of components, namely:

- estimation of flood hazard (based on the depth of flooding, typical velocity and the likely debris);
- estimation of area vulnerability (based upon the nature of the area [e.g. type of property], the speed of onset of flooding and the effectiveness of the local flood warning); and
- estimation of the vulnerability of the local population (based upon the percentage of residents aged 75 and over and those suffering from a long-term illness).

The scores for these components are combined to estimate the number of injuries and deaths. The years of life lost (YLL) resulting from the estimated fatalities were calculated by subtracting the average age of the case study area population from the average life expectancy (78.5), assuming than men and women are likely to be equally affected.

6.5.3.2 WHO methodology

In order to estimate the amount of illness resulting from a flooding event, the baseline illness rate ($IR_{baseline}$) was multiplied by the excess relative risk of illness (RR-1), assuming that those unaffected by flooding have a relative risk of illness of 1. Values for relative risk and baseline illness in the general population were derived from the literature, as outlined under each identified health impact. The derived incidence rate ($IR_{flooded}$) was then applied to the number of people affected by flooding, as follows:

$$IR_{flooded} = IR_{baseline} \times (RR - 1)$$

6.5.3.3 Summary of health impacts and quantification measures

This section summarises the health impacts that have been quantified in the health impact evaluation and the measures used to do so.

Table 6.2. Health impacts and quantification measures

Impact	Baseline incidence rate	Severity weighting	Assumed duration (days)	Relative risk
Death	Figures for deaths are calculated from the flood risks to people formula, assuming an average life expectancy of 78.5. The age of death (due to a flooding incident) is taken as the average age for those living in the area.			
Serious injury*	–	0.372	56	–
Minor injury	–	0.05	7	–
Asthma (adults only)	0.076	0.07	7	3.1
Gastrointestinal illness	0.02	0.09	3	2.36
Earache	0.001	0.09	3	2.2
Mental health (adults only)	0.155	0.065	365	4.1

* based on a broken leg

6.5.4 Results

As both case study areas only have a small number of properties that are typically flooded, relatively few people are affected and, thus, health impacts, in terms of DALYs, are correspondingly small. The results are summarised in the following sections.

6.5.4.1 Deaths and serious harm

No deaths or cases of serious harm are predicted to occur from the number of properties flooded and the depths and velocities of the water during the flooding events. This supports anecdotal knowledge from the events and also the questionnaire responses.

6.5.4.2 Physical symptoms

Table 6.3 outlines the number (or fraction) of people affected and the corresponding years lived with a disability (YLD) score for the case study areas.

Table 6.3. Physical symptoms

	Minor injuries		Asthma		GI		Earache	
Area	No.	YLD	No.	YLD	No.	YLD	No.	YLD
Utley (Hollins Lane)	1.7	0.002	3.8	0.005	0.8	0.0006	0.03	0.00002
Utley (Beechcliffe)	1.2	0.001	2.4	0.003	0.5	0.0004	0.02	0.00002
Devonshire Park	1.8	0.002	3.5	0.005	0.8	0.0006	0.03	0.00003

GI: Gastrointestinal illness YLD: Years lived with disability

6.5.4.3 Mental health problems

Mental health problems, characterised as psychological distress, were estimated for adults. Due to the long duration of these problems the YLD scores were greater than for the combined physical symptoms outlined above, with 0.582 being estimated for Utley and 0.332 estimated for Devonshire Park.

6.5.4.4 Overall health impact

Table 6.4 outlines the estimated DALYs as a result of flooding in the case study areas.

Table 6.4. Health impact of flooding

Area	Deaths (YLL)	Serious Injuries (YLD)	Other physical symptoms (YLD)	Mental health symptoms (YLD)	Total (DALYs)
Utley (Hollins Lane)	0.03	0.01	0.006	0.36	0.405
Utley (Beechcliffe)	0.03	0.01	0.005	0.22	0.264
Devonshire Park	0.05	0.02	0.007	0.33	0.407

YLL: Years of life lost YLD: Years lived with disability

Figure 6.1 shows the impact on health from a single flood event to each case study area. Although flooding is relatively rapid, the water is not moving at very high velocities, it is not very deep, and only a few properties are affected, which accounts for the lack of DALYs from deaths or serious injuries. Under these circumstances the health impacts are dominated by effects on mental health.

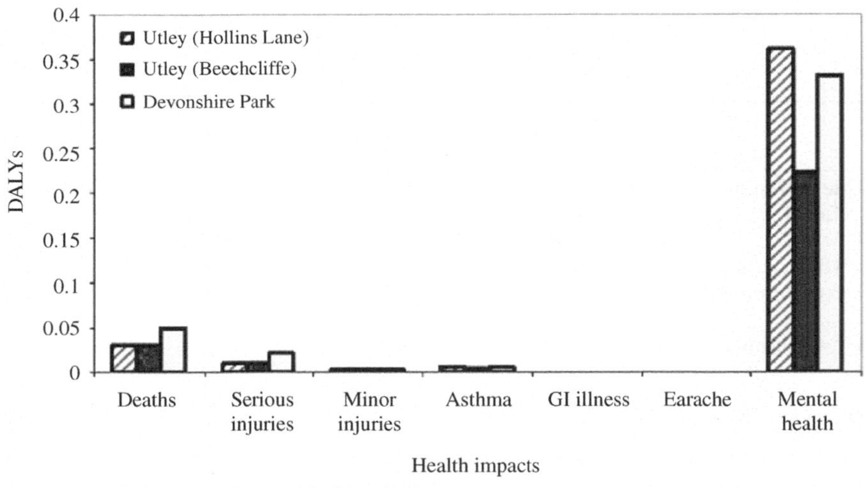

Figure 6.1. Summary of quantified health impacts

6.6 DISCUSSION

6.6.1 Vulnerability and hastened mortality

Some groups, such as pregnant women, the young, the elderly and immunocompromised people are more vulnerable to health impacts such as infection (Flynn and Nelson 1998; White *et al.* 2002; WHO 2004). It has not been possible adequately to reflect this heterogeneity within either the literature review or the case study quantifications of health impact. Although many studies note that certain groups may be more or less vulnerable to health impacts (e.g. Huerta and Horton 1978 – who actually noted that the elderly in their study coped well in terms of emotional effects), few studies have enough participants to be able to quantify the interaction between age and these effects. In addition to vulnerable groups that can, to some degree, be predicted, Few *et al.* (2004) note that "*vulnerability is shaped by a combination of physical, social, economic and environmental factors – the attributes of the person/system that condition the impacts resulting from flooding.*"

In the case study quantification, the proportion of elderly people (aged 75 and over) was considered in terms of assigning population vulnerability to estimate the number of fatalities and people experiencing serious harm (DEFRA/EA 2006), but not in determining the other health effects.

It has also been noted that, in some instances, flooding seems to hasten mortality in the elderly or chronically sick (Bennet 1970). Anecdotal evidence from flooding in Banbury and Kidlington (Tapsell 2000) supports this with a number of people being upset about elderly neighbours who had died shortly after the flooding and feeling that the experience had accelerated their demise. There are, however, very little data with which to quantify this effect and it is not accounted for within the quantification (as this only reflects deaths up to a week after the event). Deaths other than drowning are included as a heart attack during an evacuation while not caused by the water, is nevertheless directly related to the flood event (Kelman 2004).

The FHRC have developed a social flood vulnerability index, which is based upon the number of people aged 75 and over, people suffering from long term illness, lone parent households and financially deprived households (Theresa Wilson, pers. comm.). This could be employed to determine areas that may have more susceptible populations but it can not currently be translated into differential health impacts.

6.6.2 Mental health problems

A number of studies and commentaries have noted that mental health problems are one of the most significant health impacts of flooding. In a workshop evaluating the health impacts related to Hurricane Floyd in the USA for example, "*many participants felt the most egregious long-term health effects will be those related to the psychological stress associated with the storm*" (Booker 2000), despite the fact that 52 people were killed as a result of the flood (CDC 2000a).

While the immediate effects of flooding, both in terms of health impact and damage to the property, are the most obvious, Tapsell and Tunstall (2001) note that "*the aftermath of the flooding and coping with the recovery process was for the majority of participants the most stressful and difficult time*".

It is clear from this quantification of health impacts of flooding that mental health problems are the most significant, being an order of magnitude greater than the next largest impact. The size of the health impact is largely due to the long-lived nature of the effect, with people still reporting problems after a year.

6.6.3 Case studies

To our knowledge, this is the first attempt to quantify a wide range of health impacts as a result of flooding in this way. The case studies reported here were based on small areas (populations between 226 and 553) with few flooded properties (8 to 12), however, the calculations could be applied to any flooded area and used to compare different events.

There is some uncertainty about the actual number of properties flooded and no distinction was made between business and domestic properties (all the flooded properties were assumed to be domestic).

6.6.4 Lack of data and data uncertainty

6.6.4.1 Health data

As mentioned in the introduction, the health impacts of flooding do not lend themselves to controlled prospective epidemiological studies, and so much of the health information is derived from opportunistic retrospective studies of flooding. Routine surveillance data can be useful in determining episodes of infectious disease before and after flooding events, although this is, at best, a crude tool because it records only a small proportion of, for example, self-limiting gastroenteritis. Thus, it is often not possible to obtain accurate information on disease incidence or prevalence before the flood. It is likely that detection may increase after the flood (because of enhanced surveillance activities) and this variation can bias the estimates. In addition, surveillance only captures a small proportion of some illnesses and, in many cases, the connection with a flooding episode may not be made by the respondent. As a result (and as noted by Few *et al.* 2004) "*generally, there is very weak data available on non-drowning (non-immediate) deaths that can be attributed to the flood event*".

The risk values which have been derived from the literature and used in the estimation of health impacts could be site and/or flood-type specific. Although evidence suggests that the types of health impacts are similar irrespective of the type of flooding, the magnitude of health impact may well vary. For example, gastrointestinal effects could be far greater as a result of household flooding of an urban river with an upstream wastewater treatment works input compared to flooding caused by run-off from parkland.

It has been difficult to adopt a consistent approach to the estimation of health impacts, because of the different forms in which data were available in the literature for the different health impacts.

6.6.4.2 DALY quantification

There are limited data on the duration of most flood-related health impacts, therefore this information has been drawn from the general medical literature. In some cases 'indicator' conditions have been used in order to ascribe figures to both duration and severity and simplify the calculations. In the case of serious injuries, for example, these have been assumed to be equivalent to a broken leg and ascribed a severity weight and duration accordingly. Where more detailed information is available, say for a specific flood event, the actual injuries sustained could be included in the DALY calculation.

Generally, there is good information on appropriate severity weights (VGDHS 1999). Where a specific health outcome does not have a pre-defined weight this has been extrapolated from similar health outcomes.

In the case studies, relatively few people are affected by flooding, this means that, in many cases, that the calculations give fractions for the number of deaths, injuries and illnesses. Clearly it is impossible to have part of a death, but the calculations are meant to be indicative of the likely health impact of an event and, thus, DALYs have been calculated irrespective of whether integers are returned.

It has not been possible to account for all the known health impacts, due to a lack of empirical data, therefore, the total DALY score will not capture the entire health impact experienced by the community affected by a flooding event. However, the methodology presented here does allow different flood events to be compared for a number of specified impacts.

6.6.4.3 Flooding data

It is generally difficult to determine the total number of properties and people affected by a flooding event. It is certainly rare that these figures are reported in health-based studies. In the case studies reported here, questionnaire-based research was conducted some time after the flooding events in order to try to determine the scale of the flooding. Unfortunately, due to interviewing constraints (many people were at work when the interviewers called) and the time lapse (a number of people who were interviewed were new to the areas since the flooding) relatively limited information was obtained. However, the need for improved data in this area has been recognised and it is planned that Bradford Metropolitan District Council will continue this work and target questionnaires at areas as they are flooded. It has been noted, however, that there is a resistance to reporting flooding because of the perceived impact on property prices (R. Ashley, pers. com.).

6.7 ACKNOWLEDGEMENTS

Thanks to Tony Poole of Bradford Metropolitan District Council and John Blanksby of the Pennine Water Group for advice and input of local knowledge on the case study areas.

6.8 REFERENCES

Aavitsland, P., Iversen, B.G., Krigh, T., Fonahn, W. and Lystad, A. (1996) [Communicable diseases during the 1995 floods in eastern Norway]. *Tidsskrift for den Norske Laegeforening* **116**, 2038–2043. [Article in Norwegian].

Abrahams, M.J., Price, J., Whitlock, F.A. and Williams, G. (1976) The Brisbane floods, January 1974: their impact on health. *Medical Journal of Australia* **2**, 936–939.

Appelman, C.L.M., van Balen, F.A.M., van de Lisdonk, E.H., van Weert, H.C.P.M. and Eizenga, W.H. (1999) NHG Practice Guidance Acute otitis media. The Dutch College of General Practitioners Practice Guideline.

Auger, C., Latour, S., Trudel, M. and Fortin, M. (2000) [Post-traumatic stress disorder. After the flood in Saguenay]. *Canadian Family Physician* **46**(12), 2420–2427. [Article in French].

Bennet, G. (1970) Bristol floods 1968. Controlled survey of effects on health of local community disaster. *British Medical Journal* **3**, 454–458.

Bokszczanin, A. (2002) Long-term negative psychological effects of a flood on adolescents. *Polish Psychological Bulletin* **33**(1), 55–61.

Booker, S.M. (2000) Evaluating Floyd's effect on health in eastern North Carolina. *Environmental Health Perspectives* **108**(2), A67.

Buckley, A., Dawson, A., Moss, S.R., Hinsley, S.A., Bellamy, P.E. and Gould, E.A. (2003) Serological evidence of West Nile virus, Usutu virus and Sindbis virus infection of birds in the UK. *Journal of General Virology* **84**, 2807–2817.

CDC (1993) Morbidity surveillance following the Midwest flood – Missouri, 1993. *Morbidity and Mortality Weekly Report* **42**(41), 797–798.

CDC (1994a) Flood related mortality – Georgia, July 4–14, 1994. *Morbidity and Mortality Weekly Report* **43**(29), 526–530.

CDC (1994b) Acute pulmonary hemorrhage/hemosiderosis among infants – Cleveland, January 1993 – November 1994. *Morbidity and Mortality Weekly* **43**, 881–883.

CDC (1994c) Rapid assessment of vectorborne diseases during the Midwest flood – United States, 1993. *Morbidity and Mortality Weekly* **43**(26), 481–483.

CDC (2000a) Morbidity and mortality associated with Hurricane Floyd – North Caroline, September-October 1999. *Morbidity and Mortality Weekly Report* **49**(17), 369–372.

CDC (2000b) Update: Pulmonary hemorrhage/hemosiderosis among infants – Cleveland, Ohio, 1993–1996. *Morbidity and Mortality Weekly* **49**(9), 180–184.

CDC (2002) Tropical storm Allison rapid needs assessment – Houston, Texas, June 2001. *Morbidity and Mortality Weekly Report* **51**(17), 365–369.

CDC (2005) http://www.bt.cdc.gov/disasters/floods/recovery.asp (accessed 24.02.05).

CDC (2006a) Public health response to Hurricanes Katrina and Rita – Louisiana, 2005. *Morbidity and Mortality Weekly Report* **55**(2), 29–30.

CDC (2006b) Health hazard evaluation of police officers and firefighters after Hurricane Katrina – New Orleans, Louisiana, October 17–28 and November 30 – December 5, 2005. *Morbidity and Mortality Weekly Report* 55(16), 456–458.

CDR (2005a) Common animal associated infections, England and Wales laboratory reports: weeks 49–52/04. *Communicable Disease Report Weekly* **15**(6).

CDR (2005b) West Nile virus surveillance plans for 2005, and update on the 2004 season. *Communicable Disease Report Weekly* **15**(22).

Clemens, P., Hietala, J.R., Rytter, M.J., Schmidt, R.A. and Reese, D.J. (1999) Risk of domestic violence after flood impact: effects of social support, age, and history of domestic violence. *Applied Behavioural Science Review* **7**(2), 199–206.

Coker, A.L., Hanks, J.S., Eggleston, K.S., Risser, J., Tee, P.G., Chronister, K.J., Troisi, C.L., Arafat, R. and Franzini, L. (2006) Social and mental health needs assessment of Katrina evacuees. *Disaster Management and Response* **4**(3), 88–94.

Dales, R.E., Burnett, R. and Zwanenburg, H. (1991a) Adverse health effects among adults exposed to home dampness and molds. *American Review of Respiratory Diseases* **143**, 505–509.

Dales, R.E., Zwanenburg, H., Burnett, R. and Franklin, C.A. (1991b) Respiratory health effects of home dampness and molds among Canadian children. *American Journal of Epidemiology* **134**(2), 196–203.

Daley, W.R., Shireley, L. and Gilmore, R. (2001) A flood-related outbreak of carbon monoxide poisoning – Grand Forks, North Dakota. *The Journal of Emergency Medicine* **21**(3), 249–253.

Dearborn, D.G., Infield, M.D., Smith, P.G., Brooks, L.J., Carroll-Pankhurst, C., Kosick, R., Dahms, B.B., Balraj, E.K., Challener, R., Allan, T.M., Horgan, T.E., Staib, R., Wallace, C., Halpin, T.J. and Miller, J. (1997) Update: pulmonary hemorrhage/ hemosiderosis among infants – Cleveland, Ohio, 1993–1996. *Morbidity and Mortality Weekly* **46**, 33–34.

DEFRA/EA (2003) The appraisal of human-related intangible impacts of flooding. R&D Project FD2005. DEFRA/Environment Agency Flood and Coastal Defence R&D Programme. Policy Development Theme, London.

DEFRA/EA (2004) Flood risks to people, Phase 2. FD2321/IR2. DEFRA/Environment Agency Flood and Coastal Defence R&D Programme. October 2004.

DEFRA/EA (2006) Flood risks to people. Phase 2. The risks to people methodology. FD2321/TR1. DEFRA/Environment Agency Flood and Coastal Defence R&D Programme. March 2005

DoH (1999) Health Survey for England 1998. Cardiovascular Diseases, (eds B. Erens and P. Primatesta), The Stationery Office, London.

Duclos, P. and Isaacson, J. (1987) Preventable deaths related to floods. *American Journal of Public Health* **77**(11), 1474.

Duclos, P., Vidonne, O., Beuf, P., Perray, P. and Stoebner, A. (1991) Flash flood disaster – Nîmes, France, 1988. *European Journal of Epidemiology* **7**(4), 365–371.

Duke, L.A., Breathnach, A.S., Jenkins, A.S., Harris, B.A. and Codd, A.W. (1996) Mixed outbreak of *Cryptosporidium* and *Campylobacter* infection associated with a private water supply. *Epidemiology and Infection* **116**, 303–308.

Durkin, M.S., Khan, N., Davidson, L.L., Zaman, S.S. and Stein, Z.A. (1993) The effects of a natural disaster on child behaviour: evidence for posttraumatic stress. *American Journal of Public Health* 83(11), 1549–1553.

DWI (2003) Information leaflet: Private Water Supplies. Drinking Water Inspectorate. http://www.dwi.gov.uk/consumer/faq/private.htm accessed 15/07/05

European Opinion Research Group (2003) Health, food and alcohol and safety. *Special Eurobarometer* **186**.

Elidemir, O., Colasurdo, G.N., Rossmann, S.N. and Fan, L.L. (1999) Isolation of Stachybotrys from the lung of a child with pulmonary hemosiderosis. *Pediatrics* **104**, 964–966.

EM-DAT (2005) http://www.cred.be/emdat/ accessed 28.01.05

Evans, E.P., Ashley, R., Hall, J., Penning-Rowsell, E.C., Saul, A., Sayers, P., Thorne, C. and Watkinson, A. (2004) Foresight Future Flooding Scientific Summary. Volumes I and II. Office of Science and Technology, London.

Few, R., Ahern, M., Matthies, F. and Kovats, S. (2004) Floods, health and climate change: a strategic review. Tyndall Centre for Climate Change. Working Paper 63.

Fewtrell, L. (1991) Freshwater recreation: a cause for concern? *Applied Geography* **11**, 215–226.

Fewtrell, L., Kay, D. and Godfree, A. (1998) The microbiological quality of private water supplies. *Journal of the Chartered Institution of Water and Environmental Management* **12**, 98–100.

Fewtrell, L. (2002) West Nile Virus. Report to the Environment Agency. EA Project No 76.

Flynn, B.W. and Nelson, M.E. (1998) Understanding the needs of children following large-scale disasters and the role of government. *Child and Adolescent Psychiatric Clinics of North America* 7(1), 211–230.

Fuortes, L. and Nettleman, M. (1994) Leptospirosis: a consequence of the Iowa flood. *Iowa Medicine* **84**(10), 449–450.

Gallup, J., Kozak, P., Cummins, L. and Gillman, S. (1987) Indoor mold spore exposure: characteristics of 127 homes in Southern California with engogenous mold problems. *Experientia Supplementum* **51**, 139–142.

Gilliland, T.M., Rowley, W.A., Swack, N.S., Vandyk, J.K. and Bartoces, M.G. (1995) Arbovirus surveillance in Iowa, USA, during the flood of 1993. *Journal of the American Mosquito Control Association* **11**(2), 157–161.

Ginexi, E.M., Weihs, K., Simmens, S.J. and Hoyt, D.R. (2000) Natural disaster and depression: a prospective investigation of reactions to the 1993 Midwest floods. *American Journal of Community Psychology* **28**, 495–518.

Guillet, P., Germain, M.C., Giacomini, T., Chandre, F., Akogbeto, M., Faye, O., Kone, A., Manga, L. and Mouchet, J. (1998) Origin and prevention of airport malaria in France. *Tropical Medicine and International Health* **3**(9), 700–705.

Hajat, S., Ebi, K.L., Kovats, S., Menne, B., Edwards, S. and Haines, A. (2003) The human health consequences of flooding in Europe and the implications for public health: a review of the evidence. *Applied Environmental Science and Public Health* **1**(1), 13–21.

Han, L.L., Popovici, F., Alexander Jr., J.P., Laurentia, V., Tengelsen, L.A., Cernescu, C., Gary Jr., H.E., Ion-Nedwlcu, N., Campbell, G.L. and Tsai, T.F. (1999) Risk factors for West Nile virus infection and meningoencephalitis, Romania, 1996. *Journal of Infectious Diseases* **179**, 230–233.

Hardin, B.D., Kelman, B.J. and Saxon, A. (2003) Adverse human health effects associated with molds in the indoor environment. *Journal of Occupational and Environmental Medicine* **45**(5), 470–478.

Higgs, S., Snow, K. and Gould, E.A. (2004) The potential for West Nile virus to establish outside of its natural range: a consideration of potential mosquito vectors in the United Kingdom. *Transactions of the Royal Society of Tropical Medicine and Hygiene* **98**, 82–87.

Hoare, J., Bruce, M. and Majeed, A. (2003) Prevalence of treated asthma and its management in general practice in England and Wales, 1994–1998. *Health Statistics Quarterly* **17**, 15–22.

Hubalek, Z., Halouzka, J., Juricova, Z., Prikazsky, Z., Zakova, J. and Sebesta, O. (1999) [Surveillance of mosquito-borne viruses in Breclav after the flood of 1997] *Epidemiologie, mikrobiologie, imunologie : casopis Spolecnosti pro epidemiologii a mikrobiologii* **48**(3), 91–96. [Article in Czech].

Huerta, F. and Horton, R. (1978) Coping behaviour of elderly flood victims. *The Gerontologist* **18**(6), 541–546.

Kelman, I. (2003a) CURBE Fact Sheet 3: UK Deaths from the 1953 storm surge. Version 4, 10 November 2003.

Kelman, I. (2003b) UK Drownings. CFS Press, January 2003.

Kelman, I. (2004) Philosophy of flood fatalities. http://www.floodrisknet.org.uk/newsletters/2004-01/flood-fatalities

Kibble, A., Pooransingh, S. and Saunders, P. (2003) *Stachybotrys* a growing threat? *Strategic Risk* February 2003.

Kool, J.L., Warwick, M.C., Pruckler, J.M., Brown, E.W. and Butler, J.C. (1998) Outbreak of Legionnaires' disease at a bar after basement flooding. *Lancet* **351**, 1030.

Kuhn, D.M. and Ghannoum, M.A. (2003) Indoor mold, toxigenic fungi, and Stachybotrys chartarum: infectious disease perspective. *Clinical Microbiology Reviews* **16**(1), 144–172.

Leal-Castellanos, C.B., Garcia-Suarez, R., Gonzalez-Figueroa, E., Fuentes-Allen, J.L. and Escobedo-de la Penal, J. (2003) Risk factors and the prevalence of leptospirosis infection in a rural community of Chiapas, Mexico. *Epidemiology and Infection* **131**(3), 1149–1156.

McMillen, C., North, C., Mosley, M. and Smith, E. (2002) Untangling the psychiatric comorbidity of posttraumatic stress disorder in a sample of flood survivors. *Comprehensive Psychiatry* **43**(6), 478–485.

Medlock, J.M., Snow, K.R. and Leach, S. (2005) Potential transmission of West Nile virus in the British Isles: an ecological review of candidate mosquito bridge vectors. *Medical and Veterinary Entomology* **19**, 2–21.

Milly, P.C., Wetherald, R.T., Dunne, K.A. and Delworth, T.L. (2002) Increasing risk of great floods in a changing climate. *Nature* **415**, 514–517.

Mouchet, J. (2000) Airport malaria: a rare disease still poorly understood. *Eurosurveillance* **7**(7/8),75–76.

Ogden, C.L., Gibbs-Scharf, L.I., Kohn, M.A. and Malilay, J. (2001) Emergency health surveillance after severe flooding in Louisiana, 1995. *Prehospital and Disaster Medicine* **16**(3), 138–144.

Ohl, C. and Tapsell, S. (2000) Flooding and human health. *British Medical Journal* **321**, 1167–1168.

Payment, P. and Hunter, P.R. (2001) Endemic and epidemic infectious intestinal disease and its relationship to drinking water. In *Water Quality: Guidelines, Standards and Health* (eds L. Fewtrell and J. Bartram), pp. 61–88, IWA Publishing, London.

Petrie, A.S., Horan, N.J., Clapham, D.B. and Cram, A.G. (1994) Seasonal variation in the quality of spring water used as private supplies. *Journal of the Institute of Water and Environmental Management* **8**, 320–326.

Prüss, A., Kay, D., Fewtrell, L. and Bartram, J. (2002) Estimating the burden of disease from water, sanitation, and hygiene at a global level. *Environmental Health Perspectives* **110**(5), 537–542.

Reacher, M., McKenzie, K., Lane, C., Nichols, T., Kedge, I., Iversen, A., Hepple, P., Walter, T., Laxton, C. and Simpson, J. (2004) Health impacts of flooding in Lewes: a comparison of reported gastrointestinal and other illness and mental health in flooded and non-flooded households. *Communicable Disease and Public Health* **7**(1), 39–46.

Riddez, L. and Dellgar, U. (2005) KAMEDO Report No. 76: Floods in Poland in 1997 and in Sweden in 2000. *Prehospital and Disaster Medicine* **20**(4), 262–264.

Robson, A.J. (2002) Evidence for trends in UK flooding. *Philosophical transactions of the Royal Society of London. Series A: Mathematical and physical sciences* **360**, 1327–1343

Spice, B. (2004) Infection from floodwater fatal to firefighter. *Pittsburgh Post-Gazette* 28.09.04.

Sullivent, E.E. 3rd, West, CA., Noe, R.S., Thomas, K.E., Wallace, L.J. and Leeb, R.T. (2006) Nonfatal injuries following Hurricane Katrina – New Orleans, Louisiana, 2005. *Journal of Safety Research* **37**(2), 213–217.

Tapsell, S.M. (2000) The health effects of flooding. A follow-up study of the 1998 Easter flooding in Banbury and Kidlington. Report to the Environment Agency. Flood Hazard Research Centre, Middlesex.

Tapsell, S.M. and Tunstall, S.M. (2001) The health and social effects of the June 2000 flooding in the North East Region. Report to the Environment Agency. Flood Hazard Research Centre, Middlesex.

Tapsell, S. and Tunstall, S. (2003) An examination of the health effects of flooding in the United Kingdom. *Journal of Meteorology* **28**, 341–349.

Tapsell, S., Penning-Rowsell, E.C., Tunstall, S.M. and Wilson, T.L. (2002) Vulnerability to flooding: health and social dimensions. *Philosophical Transactions of the Royal Society of London, Series A* **360**, 1511–1525.

Tunstall, S., Tapsell, S., Green, C., Floyd, P. and George, C. (2006) The health effects of flooding: social research results from England and Wales. *Journal of Water and Health* **4**(3), 365–380.

Verger, P., Hunault, C., Rotily, M. and Baruffol, E. (2000) [Risk factors for post traumatic stress symptoms five years after the flood in the Vaucluse (France)] *Revue Epidemiol Sante Publique* **48**(Suppl 2), 2S44–2S53. [Article in French]

Verger, P., Rotily, M., Baruffol, E., Boulanger, N., Vial, M., Sydor, G., Pirard, P. and Bard, D. (1999) [Evaluation of the psychological consequences of environmental catastrophes: a feasibility study based on the 1992 floods in Vaucluse] *Sante* **9**(5), 313–318. [Article in French].

VGDHS (1999) *Victoria Burden of Disease Study: Morbidity*. Victorian Government Department of Human Services, Melbourne, Victoria, Australia.

von Wageningen, N., Waegenmaekers, M., Brunekreef, B. and Boleij, J. (1987) Health complaints and indoor moulds in relation to moist problems in homes. In *Proceedings of the Fourth International Conference on Indoor Air Quality and Climate* (eds B. Seifert, H. Esdorn, M. Fischer, H. Riiden and J. Wegner), pp. 723–727, Institute for Water, Soil and Air Hygiene, Berlin.

Wade, T.J., Sandhu, S.K., Levy, D., Lee, S., LeChevallier, M.W., Katz, L. and Colford Jr, J.M. (2004) Did a severe flood in the Midwest cause an increase in the incidence of gastrointestinal symptoms? *American Journal of Epidemiology* **159**, 398–405.

White, S.R., Henretig, F.M. and Dukes, R.G. (2002) Medical management of vulnerable populations and co-morbid conditions of victims of bioterrorism. *Emergency Medicine Clinics of North America* **20**(2), 365–392.

WHO (2004) Guidelines for Drinking-water Quality. Third edition. Volume 1 Recommendations. World Health Organization, Geneva.

WHO (2005) Flooding and Communicable Disease Fact Sheet. Risk assessment and preventative measure. World Health Organization, Geneva, http://www.who.int/hac/techguidance/ems/flood_cds/en/ accessed 24.02.05

Zitek, K. and Benes, C. (2005) [Longitudinal epidemiology of leptospirosis in the Czech Republic (1963–2003)]. *Epidemiologie, Mikrobiologie, Imunologie* **54**(1), 21–26. [Article in Czech].

7

An exploratory quantitative microbial risk assessment of urban flooding and gastrointestinal illness

Lorna Fewtrell, Keren Smith, John Watkins, John Blanksby and Adrian Saul

7.1 INTRODUCTION

The overall evaluation of the health impacts of flooding in the UK, based on risk values derived from the literature, was examined in Chapter 6. One of the areas, highlighted in that examination, was the uncertainty in relation to gastro-intestinal infection resulting from flooding, although this health concern is often picked up in media reports (Anon 2007). The work reported in this chapter applies quantitative microbial risk assessment (QMRA) to examine gastro-intestinal infection risks from flooding in a case study area.

 Health Impact Assessment for Sustainable Water Management. Edited by Lorna Fewtrell and David Kay. ISBN: 9781843391333. Published by IWA Publishing, London, UK.

For there to be an impact of flooding on human health a number of factors have to coincide, these can be thought of in terms of Source, Pathway and Receptor, as illustrated in Figure 7.1. Where, in simple terms, the Source is the microbially polluted floodwater, the Pathway is the route that the floodwater takes and the Receptor is the householder affected by the floods.

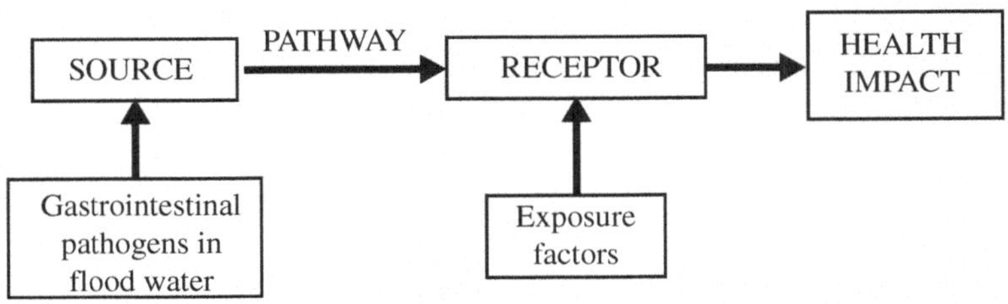

Figure 7.1. Source, pathway, receptor model and health consequences

There are a number of component flood flows which may contribute to the 'quality' of the floodwater, such as:

- foul flow (including combined sewer overflows – CSOs);
- urban non-foul pollution (including roof run-off); and
- rivers (including rural diffuse pollution).

The microbial loading of these possible flood flows was investigated through an extensive literature review (Smith *et al.* 2007), opportunistic sampling and experimental work in order to inform input parameters for a QMRA.

There are far too many pathogens (and a paucity of data about most) to consider all possible causes of infection that could be present within floodwater and sediment. A common approach is to consider a number of reference pathogens, usually consisting of a bacterial, viral and protozoan pathogen (WHO 2004). Suitable reference pathogens are usually those that present a worst case combination (NRMMC 2006) of high occurrence, relatively high concentrations in key flood components, high pathogenicity and environmental survival.

Based on an examination of the literature relating to floodwater quality (Smith *et al.* 2007), the most frequently identified bacterial and protozoan pathogens are *Campylobacter* spp., *Salmonella* spp., *Cryptosporidium* spp. and *Giardia* spp. Campylobacters are the most commonly reported bacterial cause of bacterial gastroenteritis in the UK, with an estimated incidence of 8.7/1000 population (Adak *et al.* 2002) and on this basis were chosen as the bacterial representative. *Cryptosporidium* infection is more prevalent than *Giardia* infection (estimated incidence in the UK 8.1/10,000 and 5.4/10,000 respectively – Adak *et al.* 2002). *Cryptosporidium* has also been more frequently associated with waterborne illness than *Giardia* in the UK, and was chosen as the protozoan representative.

The choice for virus is less clear cut. Noroviruses are the most common cause of viral gastroenteritis in the UK, with an estimated annual incidence rate of 12.1/1000 (Adak *et al.* 2002) but there is no dose-response relationship for this virus. Rotaviruses are the second most common cause of viral gastroenteritis and a dose-response relationship has been established (Haas and Eisenberg 2001), but concentrations can not be determined by cell culture. A composite virus was, therefore, chosen, based on concentrations of adenovirus (which can be determined by cell culture), but with the dose-response characteristics of rotavirus.

7.2 MICROBIAL CONTAMINATION OF FLOOD COMPONENTS

7.2.1 Sewage

Raw (untreated) sewage may become a flood component in a number of ways, such as where a sewer becomes blocked, where the sewerage system is overwhelmed or where a combined sewer overflow discharges. The microbial contamination will vary both temporally and spatially according to the health and the size of the contributing population. CSOs are also likely to be a significant source of faecal indicator organisms and pathogens to urban waters in the UK, although data is very limited.

Treated wastewater may also be a component of flood flow through discharges from wastewater treatment works (WWTW) to receiving waters, or through an overflow related to heavy rainfall. Concentrations will vary according to the quality of the influent sewage, and the nature and extent of the treatment processes. There is much evidence that, despite treatment, significant concentrations of pathogens may be present in WWTW effluent, although UK data is sparse (Kay *et al.* 2008). In terms of faecal indicator organisms, following efficient treatment, wastewater may have concentrations far lower than those observed in stormwater, urban runoff and the surface waters to which the effluent is discharged.

7.2.1.1 Literature values

Relatively few published studies have isolated or enumerated pathogens in raw sewage in the UK, although there is evidence of potentially high concentrations.

Campylobacter spp. have been reported in UK sewage at a level of 10^5/100ml (Arimi *et al.* 1988). Outside the UK, research has also identified significant concentrations of *Salmonella* spp. and *Shigella* spp., as well as many pathogenic viruses, and both *Cryptosporidium* and *Giardia,* with up to 100% prevalence. Research in the UK on faecal indicator organism concentrations in raw sewage indicates faecal coliform concentrations in the order of 10^6–10^7 cfu/100ml and intestinal enterococci at 10^5–10^6 cfu/100ml (Kay *et al.* 2008). The sections below represent a summary of the literature review in relation to the pathogens chosen for the QMRA, with a focus (where possible) on UK data.

Campylobacter spp.

Arimi *et al.* (1988) investigated the occurrence of campylobacters in sewage from a wastewater treatment works (WWTW) in Reading, UK. Levels in incoming sewage (i.e. raw sewage) were found to fluctuate, with a maximum recorded concentration of 4.6×10^5 MPN/l. In treated sewage mean concentrations were approximately 100 MPN/l.

Cryptosporidium spp.

In the UK, Wyer *et al.* (2001) found *Cryptosporidium* spp. in all 21 samples of crude sewage effluent taken from a works in the Irvine catchment, Scotland. A significant elevation in concentration was found at high flows, with a geometric mean concentration of 55.1 oocysts/litre, compared with 17.6 oocysts/litre at baseflow. In the USA, Rose (2007) reported that *Cryptosporidium* concentrations in untreated sewage ranged from 10^1 to 10^4 oocysts per 100l.

No data on *Cryptosporidium* concentration in treated effluent from the UK were located. The extent of wastewater treatment influences the concentrations remaining in final effluent. In a study of ten WWTWs in the USA, McCuin and Clancy (2006) detected *Cryptosporidium* oocysts in 46% of primary, 58% of secondary, and 19% of tertiary effluent samples. In five WWTWs in Spain, Montemayor *et al.* (2005) found *Cryptosporidium* oocysts in 100% of 24 secondary-treated samples (concentrations 0.4–16 oocysts/l, with viability between 0–50%), compared with 70% of tertiary-treated samples (concentrations 0.2–0.8 oocysts/l, with viability between 0–25%).

Because of the component of raw sewage present in a CSO, the occurrence and concentrations of *Cryptosporidium* spp. are expected to be of the order of those found in untreated effluent. Arnone and Walling (2006) detected *Cryptosporidium* spp. in 12% of CSO samples from three urban areas in the USA, with concentrations between 59–100 oocysts/100l. States *et al.* (1997) found the pathogen in four of five samples taken over ten months from a CSO in Pittsburgh,

USA, with concentrations ranging from 875–3000 oocysts/100l. Gibson *et al.* (1998) found 100% prevalence in a CSO discharge to Saw Mill Run Stream in the USA, on eleven occasions over five rainfall events, with a range of concentrations between 250–40,000 oocysts/100l.

Viruses

No data relating to rotavirus or adenovirus concentrations in raw or treated sewage from the UK were identified. Studies from other countries (Krikelis *et al.* 1985; Myrmel *et al.* 2006), however, suggest that adenovirus is likely to be found in raw sewage (prevalence of 96%+), often at high concentrations (70–3200 cpu/l). Rotavirus was found in raw sewage samples less frequently than adenovirus (approximately 60%), which may reflect its more seasonal distribution. Both adenovirus and rotavirus were also detected in treated sewage samples (Myrmel *et al.* 2006).

7.2.1.2 Experimental values

A series of microcosms were set up in the laboratory to study the concentration and survival of bacterial and protozoan pathogens in sewage samples. Ten litre samples of wastewater were collected from Esholt WWTW in Yorkshire on six separate occasions (in spring and autumn 2007). Samples were analysed for both faecal indicators and selected pathogens (*Salmonella* spp., *Campylobacter* spp., *E. coli* O157, *Cryptosporidium* spp. and *Giardia* spp.). All organisms were enumerated (rather than simply determining presence or absence), and analyses were conducted on both the overlying water (supernatant) and the sediment. Analyses were conducted on up to six different days; day 0, 1, 3, 7, 14 and 21. Table 7.1 summarises the range of data for *Campylobacter* spp. and *Cryptosporidium* spp.

Table 7.1. Summary of sewage microcosm data

	Campylobacter spp./litre		*Cryptosporidium* spp./litre	
Day	Supernatant	Sediment	Supernatant	Sediment
0	<10 – >180,000	–	1 – 96	–
1	0 – >180,000	160 – >180,000	0 – 12	50 – 500
3	0 – 130,000	0 – 130,000	0 – 8	100 – 700
7	0 – 750	0 – 2400	<1 – 12.5	200 – 600
14	0 – 40	0 – 10	<1 – 16	<100 – 1500
21	0 – <10	0 – 10	0 – 2	<50 – 1040

It can be seen from Table 7.1 that there is huge variability in the initial concentration of *Campylobacter* spp. in both the supernatant and sediment fractions, with relatively rapid die off during storage. There is markedly less variability in *Cryptosporidium* spp. concentration and an apparently slower die off rate (especially in sediment).

7.2.2 Urban non foul

Non-point source urban runoff in the surface drainage system includes overland runoff from impervious surfaces (including roofs), as well as misconnections from foul sewer systems. Pathogens including *Campylobacter* spp., *Salmonella* spp., *Cryptosporidium* spp. and *Giardia* spp. have been reported in roof runoff, although with infrequent isolations (Fewtrell and Kay 2007). There are few data on pathogen concentrations in urban stormwater, and the majority of studies have examined faecal indicator concentrations. Published reports examining urban (non-foul) surface runoff generally do not describe, or do not test for sanitary cross connections to the storm sewer systems, or are concerned with the effects of stormwater outfalls on riverine water quality, with inferential analysis of stormwater microbial loads and concentrations (Olivieri *et al.* 1977; Schillinger and Gannon 1985; Gannon and Busse 1989). The only pathogen analyses identified from the literature relate to *Pseudomonas aeruginosa* (Pitt and McLean 1986; Selvakumar and Borst 2006).

7.2.2.1 Literature values

Campylobacter spp.

Several studies of pathogen concentrations in roof runoff have failed to find *Campylobacter* spp. (Holländer *et al.* 1996; Simmons *et al.* 2001; Birks *et al.* 2004). The highest prevalence reported is 37.5% from roof run off samples in rural New Zealand, with a maximum concentration of 0.56 /100ml in rainwater tank storage (Savill *et al.* 2001).

Cryptosporidium spp.

Birks *et al.* (2004) did not find *Cryptosporidium* spp. in roof runoff samples in the UK (Millennium Dome). In Denmark, Albrechtsen (2002) reported *Cryptosporidium* in 35% of samples taken from rainwater harvesting tanks, with a maximum concentration of 50 oocysts/litre. This figure seems to be very high and it has been speculated that it is likely that the *Cryptosporidium* species present are most likely to be those infective in birds (*C. baileyi* and *C. meleagridis*) which are not particularly infective to humans (Fewtrell and Kay 2007).

Viruses

Human viruses would not be expected within rainwater or roof runoff, nor would they be expected in urban runoff in the absence of cross connections with the foul system. Accordingly, no data were identified.

7.2.3 Rivers

Rivers receive microbial pollution from each of the identified flood components, and can carry a wide range of pathogens including bacteria, viruses and protozoa. In UK rivers, research has identified bacterial pathogens *Salmonella* spp. and *Campylobacter* spp. No published data on viral concentrations in UK rivers were identified, although evidence from research in the Netherlands and the USA indicates the occurrence of adenoviruses, enteroviruses, norovirus, astrovirus, rotaviruses, and reoviruses, possibly in high concentrations. In terms of protozoa, *Cryptosporidium* has been isolated from rural and urban rivers in the UK, and other data indicates that rivers may have high concentrations of both *Cryptosporidium* and *Giardia* with significant viability.

In urban areas, the rural diffuse microbial component is manifested primarily as an upstream boundary condition and can be an important source of microorganisms downstream, especially in high flow scenarios. Stormwater runoff (overland and sub-surface) from pasture and farmyards can contain high concentrations of faecal microorganisms and can be a significant source of pathogens to watercourses. Direct deposition of faecal material by cattle in-stream and sediment remobilisation from rural headwaters with the onset of high flows can also provide a major source of faecal indicator organisms downstream.

Wildlife, farm stock and domestic animals are the primary diffuse sources of pathogens and faecal indicator organisms in rural areas. High concentrations of *Cryptosporidium* and *Giardia* have been reported in faecal material deposited by farm animals including pigs and poultry, and reovirus has been reported in cattle faeces. In the UK, waterfowl have been found to be a significant source of *Campylobacter* spp.

It should be noted that microbial flux in rural river systems is highly episodic and seasonal in temperate latitudes with almost all the load being moved during high flow events and a summer peak in concentration (Kay *et al.* 2007; Rodgers *et al.* 2003). In systems dominated by anthropogenic point source inputs (likely to be the case in urban rivers), this loading pattern reverses with treated effluents producing significant flux under base flow conditions (see for example Stapleton *et al.* 2008 and Crowther *et al.* 2003). A process explanation for the episodic flux pattern, seen in terms of faecal indicator organisms, is given in Wilkinson *et al.* (2006) and the area is reviewed by Kay *et al.* (2007). Similar patterns are likely for pathogens and this must be born in mind as a caveat in judging the relevance of any microbial pathogen 'spot' determinations in rivers which are difficult to interpret without associated flow information. In addition to the catchment hydrodynamic processes producing episodic and seasonal patterns, pathogens will only be detected when they are being excreted, i.e. when there is an infection in the contributing animal or human population. Again, seasonal patterns of infection are common, for example, norovirus infection is often termed winter vomiting syndrome for obvious reasons. To date, there have been very few catchment-wide studies on pathogen dynamics which have sought to apply hydrological, epidemiological and geomorphic principles to protocol design or data interpretation (see Ferguson 2005 and Ferguson *et al.* 2003, for rare exceptions where pathogen data are properly contextualised in a catchment hydrodynamic paradigm).

7.2.3.1 Literature values

Campylobacter spp.

Five studies examining the prevalence or concentration of campylobacters in UK rivers were identified (Bolton *et al.* 1987; Jones *et al.* 1990; Obiri-Danso and Jones 1999; Moore *et al.* 2001; Bates and Phillips 2005). The prevalence was found to range between 0 and 96%, with a maximum recorded value of 330 MPN/100ml (Jones *et al.* 1990). Campylobacters are more likely to be found downstream of agricultural land, especially where livestock are kept (Jones 2001). Bates and Phillips (2005) reported a significant increase in riverine *Campylobacter* spp. concentrations after inlets from ditches draining livestock farms, with concentrations of up to 100 cfu/100ml recorded during winter.

Cryptosporidium spp.

Wyer *et al.* (2000) found *Cryptosporidium* spp. concentrations in the Ystwyth catchment in Wales of up to 50 oocysts/litre following rainfall and in association with high livestock levels. In another predominantly rural catchment, the Irvine, in Scotland, Wyer *et al.* (2001) found 74% of river samples positive for the pathogen, with a broad trend of increased incidence and concentration at high flows. Maximum concentrations were recorded during hydrograph events at 59.1 oocysts/litre.

Viruses

No data on rotavirus or adenovirus concentrations were located for UK rivers. Adenovirus has been reported in Spain with a prevalence in one river of 65%

(Pina *et al.* 1998), although this was determined by a method which doesn't ascertain viability. In the USA, an adenovirus prevalence of 48.5% (37.9% infectious – i.e. capable of growth) was reported for surface waters (Chapron *et al.* 2000). In urban rivers in Japan, Tani *et al.* (1995) found an average concentration of 2.4 pfu/l, with a maximum recorded concentration of 25 pfu/l, using cell culture techniques.

7.2.3.2 Opportunistic sampling

Very heavy rainfall across the UK on July 20th, 2007, triggered a series of floods across wide areas of central and western England over the following four days. Localised (mostly urban) flash floods and extreme flows in small responsive catchments, led to subsequent extensive inundations of the Severn, Avon, Great Ouse and Thames floodplains (CEH 2007). A week after the flooding limited floodwater sampling was conducted to get a 'feel' for levels of microbial contamination.

The sampling took place on Sunday July 29th, 2007 and included locations in Worcester, Gloucester, Tewkesbury and Upton upon Severn, where the flooding had predominantly been caused by river flooding, with sewage input. A total of five floodwater samples were collected, three of which were from external locations, and analysed for faecal indicators, *Salmonella* spp., *Cryptosporidium* spp. and *Giardia* spp. Table 7.2 summarises the results.

Table 7.2. Opportunistic floodwater sampling

Sample location	*E. coli* /100ml	Enterococci /100ml	*Salmonella* /litre	*Crypto* /litre	*Giardia* /litre
Worcester racecourse	1.2×10^4	1300	0	–	–
Gloucester – parkland	100	0	1	0	0
Longford – courtyard	1000	500	0	0.1	0.2
Tewkesbury pub cellar	1.4×10^6	17,000	2	0	0.1
Upton upon Severn pub cellar	5.0×10^6	400,000	<18	–	–

– not tested – inadequate sample

It can be seen from Table 7.2 that microbial concentrations are quite varied, although in most cases the pathogen levels are low. It is clear from the high indicator concentrations that the locations of the final two samples (both from pub cellars protected from UV irradiance) allowed greater microbial survival.

7.3 EXPOSURE

For people to be infected by pathogens resulting from a flood incident they need to be exposed to the floodwater. It is likely that this will principally occur in two ways; during flood evaluation (where immersion is a distinct possibility) and during the flood clean-up process.

7.3.1 Evacuation behaviour

Because of the diverse nature of floods and flood-warning scenarios, it is difficult to make generalisations about the behavioural responses during a flood event. From research into responses to UK flooding, Tapsell (pers. comm., 2007) attributes behavioural responses according to the type of flood, length of warning and any previous flooding experience, with decisions based on obvious risks to self or family, and the perceived likely duration of flooding.

Risks are increased with exposure to the water. Residents of ground-floor only accommodation are likely to perceive a higher risk, and be more inclined to evacuate. Where higher storeys above the flood level are accessible, residents may choose to remain. During the 1998 floods in Banbury and Kidlington, Tapsell and Tunstall (2000) reported that many of those living in two-storey accommodation chose to relocate upstairs but remain in the property. However, the duration of this flood was relatively short, with most homes being flooded for around 26 hours. During severe but slow-onset riverine flooding across Europe in the 1990s, the self-regulatory attitude of the public caused problems where people refused to evacuate because they felt adequately protected in the upper storeys. In some cases people were uncooperative with evacuation efforts for so long that emergency rescues had to be undertaken under much more risky circumstances (Rosenthal and Bezuyen 2000). Flood advisors recommend the disabling of services in flooded properties, including gas and electricity supplies, and for floods of any duration above one day it is assumed that residents will eventually attempt evacuation.

Where a flood warning is given sufficiently far in advance, from official sources, and is believed by a population, it is probable that some residents will evacuate before the onset of flooding to the property. Families, especially those with small children, are assumed more likely to evacuate early, and will thus not be exposed to the floodwaters. This may also be true of residents away from home, who are subsequently prevented (either by circumstance or the intervention of the emergency services) from returning until the flood has subsided.

In some instances, flood warnings may not be issued to allow early evacuation. In the case of sudden onset flash flooding, critical levels may be reached too quickly to allow evacuation by timely warning. Relatively impermeable, urban environments are particularly vulnerable to flash flooding, and warning lead times may be as little as an hour where meteorological and environmental conditions promote rapid runoff (Gruntfest and Ripps 2000).

Even with slower-onset riverine flooding, authorities may not be sufficiently experienced, prepared or organised to issue such warnings effectively. Following extensive flooding in the Cherwell catchment in England in 1998, flood victims complained that they had not received an official flood warning (Tapsell and Tunstall 2000). The effectiveness of such warnings is unclear in any case. Evidence suggests that the initial response to a flood warning is usually threat denial, and adaptive responses are often delayed (Drabek 2000; Rosenthal and Bezuyen 2000). The effectiveness of such warnings appears to be related to previous flooding experience (Parker 1991), and may be of limited use where a population is able to dismiss the threat. Residents and emergency services without previous experience may be surprised by the speed, scope and intensity of flooding (van Duin and Bezuyen 2000; Rosenthal and Bezuyen 2000; Tapsell and Tunstall 2000). Those familiar with a severe flood may be less willing to take risks on a subsequent occasion (van Duin and Bezuyen 2000).

7.3.2 Depth of flooding and likelihood of immersion

Immersion in floodwater during evacuation may occur as a result of instability caused by a combination of water depth and velocity, or due to an unseen underwater obstacle such as a kerb, hole, or item of debris. Norwich Union (2007) suggested that six inches of fast flowing water can knock a person from their feet. Lind *et al.* (2004) and DEFRA/EA (2006) have both considered human instability in flood as a function of floodwater depth and velocity, using experimental data from Abt *et al.* (1989) and Karvonen *et al.* (2000). The product of depth and velocity at which a person becomes unstable was found to be in the range 0.65–2.13 m^2/sec (mean 1.22 m^2/sec, Lind *et al.* 2004). It is assumed in this example that the floodwater velocity in and around the house is relatively low and that, as such, human instability would be primarily a function of depth. Therefore, instability (and subsequent immersion) may be reached at flood depths above 0.65m. Immersion in shallower water is also possible given the increase in instability through motion of evacuees. Unfortunately, little information on flood depths in residential areas in the UK is available although it is clear, from anecdotal evidence reported and recorded by the media during recent flooding across the Midlands, that waters levels in excess of 0.65m occurred. Tapsell and Tunstall (2000) report flood depths from the 1998 floods in Banbury and Kidlington, UK, in most properties ranging between 0.3–0.6m.

Immersion after tripping on an unseen obstacle in floodwater has also been reported. Records from a sample of 16–18 UK hospitals kept by the Royal Society for the Prevention of Accidents (RoSPA; pers comm.) show 11 incidents of falls or trips in floodwater resulting in a hospital visit between 2000–2002 in people aged between 11–82 (with these limited data, no particular age trend is apparent).

7.3.3 Volume ingested on immersion

Assuming that the flooded population will not voluntarily submerge themselves in floodwaters for recreational purposes, immersions are likely to be brief. There are no published data available concerning volumes of water ingested in flood situations, although for the purposes of QMRA Westrell *et al.* (2004) have assumed an ingested volume per exposure of 30 ml from (un)intentional immersion.

7.3.4 Flood clean-up

Throughout the duration of the remediation process, those involved will come into contact with floodwaters, small quantities of which may be ingested accidentally through contact between hands and mouths. Tulve *et al.* (2002) report a mean frequency of mouthing behaviour in children less than 24 months of age of 81 events/hour (standard deviation 7). In older children (aged between 2–5 years) the mean frequency was significantly less, at 42 events/hour (standard deviation 4). In older children and adults this figure would be expected to be further reduced. Tulve *et al.* (2002) further report incidences of specific hand-to-mouth activity, with a mean event frequency of 18/hour in children aged 0–2 (median 12, 95% CI 9–16) and 16/hour in those aged 3–5 (median 9, 95% CI 7–12). No data on regular hand-to-mouth activity in adults has been found in the literature, although it is possible that, when exposed to floodwater, a person may attempt to minimise ingestion, especially by wearing gloves to reduce contact with the floodwater and sediment during clean-up.

If alternative arrangements can be made, it is probable that children will remain away from the property until it has been cleaned, and that chiefly adults will be exposed during this period.

The degree to which people use personal protection (gloves) during the process, and limit the transfer of water and sediments to the mouth, is likely to vary widely, and depend on the availability of such items at home, and the awareness of risk generated from the media and neighbourhood sources (Tapsell, pers. comm. 2007).

7.4 QMRA

7.4.1 Hazard characterization

The reference pathogens are outlined earlier (in section 7.1). This Section summarises the key characteristics of the pathogens in terms of QMRA and DALY quantification. Where pathogens are predicted to result in fatal cases of illness the years of life lost have been based on UK specific figures (rather than those taken from the literature used in Chapters 3 and 5).

7.4.1.1 Campylobacter spp.

Infection with *Campylobacter* spp. proceeds to clinical illness in 30% of cases (WHO 2004). The severity weight (0.086) and duration (6 days) for uncomplicated cases of campylobacteriosis are based on values described by Havelaar *et al.* (2000a) for illness in the general population and illness reported to general practitioners. The severity weight (0.28) and duration (365 days) for complicated campylobacteriosis is based on the mean value for Guillain-Barré syndrome (adapted from Havelaar *et al.* 2000a), with the incidence based on the campylobacteriosis hospitalisation rate of 0.5% reported by Mead *et al.* 1999. Recent mortality statistics (2001–2005) from the UK indicate that fatalities from campylobacteriosis occur in people aged over 65. An age-related case-fatality rate of 0.0083% for the UK has been estimated, with a mean age of death of 80 (ONS 2007) and a resultant loss of 8.3 years of life (based on additional life expectancies for males and females aged 80 of 7.57 years and 9.03 years respectively; GAD 2007).

7.4.1.2 Cryptosporidium spp.

Infection with *Cryptosporidium* spp. in developed countries is believed to result in infections in the immunocompetent population in 71% of cases, while infection in the immunocompromised population is thought to lead to illness in virtually 100% of cases (Havelaar *et al.* 2000b). The severity weights and illness duration are shown in Table 7.3 A case-fatality rate of 0.0158% has been estimated from age-related UK data on incidence, known under-reporting of non-fatal cases and population statistics (FSA 2000; Adak *et al.* 2002; HPA 2007a; ONS 2007). The mean age at death from cryptosporidiosis is 63.7 years (although for simplicity in terms of population calculations this figure was assumed to be 65).

Table 7.3. Severity weights and duration of cryptosporidiosis (Havelaar *et al.* 2000b)

Population	Severity	Duration (days)
Normal (immunocompetent)	0.054	6
Immunocompromised	0.13	47

The effect of HIV/AIDS on mortality from cryptosporidiosis infection is unclear. A CDC study in 1986 showed that cryptosporidiosis in patients with HIV/AIDS had a case fatality-rate which was significantly higher ($P < 0.01$) than the case-fatality rate for patients without reported HIV/AIDS (Navin and Hardy 1987). Furthermore, Rusin *et al.* (2000) suggested that about 10–15% of AIDS patients die of complications related to cryptosporidiosis. However, examination of UK mortality data (ONS 2007) indicates a mean age of death from cryptosporidiosis of 63.7 years, and of death from HIV disease of 42 years. Over the period 2001–2005 in the UK there were limited (6) deaths from cryptosporidiosis, and only one incident of mortality under 50 years of age. With such limited data, it is not appropriate to try to quantify mortality from cryptosporidiosis in AIDS cases separately from that of the rest of the population. In the QMRA, the influence of HIV/AIDS is thus restricted to an estimation of the additional severity and duration of the infection in the population aged 15 and over.

7.4.1.3 Viruses

Infection is assumed to proceed to clinical illness in 50% of cases (WHO 2004). The severity weight of 0.093 is based on an uncomplicated case of diarrhoea (VGDHS 1999). The duration of symptoms (vomiting and diarrhoea) are usually between 3 and 8 days (Rusin *et al.* 2000; HPA 2007b). Rotavirus infection is not usually fatal in people over the age of four (ONS 2007), thus fatalities have only been estimated for children under the age of five. The mean age at death is 1.1 years (ONS 2007). A case-fatality rate of 0.0013% has been estimated based on age-related UK data on incidence, likely under-reporting of non-fatal cases, mortality and population statistics (FSA 2000; Adak *et al.* 2002; HPA 2007b; ONS 2007).

7.4.2 Dose-response assessment

The dose-response relationships shown in Table 7.4 were used in the QMRA.

Table 7.4. Dose-response relationships

	Beta-poisson		Exponential	Reference
Pathogen	α	β	γ	
Campylobacter spp.	0.145	7.584		Medema *et al.* 1996
Cryptosporidium parvum			0.004005	Teunis *et al.* 1996
Rotavirus	0.265	0.442		Haas *et al.* 1993

7.4.3 Exposure

The demographic profile of the flooded population will obviously be case-specific. For the purposes of the exploratory QMRA, it has been assumed that 400 houses, with an occupancy level of 2.57, have been flooded. The population characteristics are based on an urban population in the north of England, using data from the 2001 census (Table 7.5).

Table 7.5 Age distribution of flooded population

	People	
Age (years)	Percentage	Number
0–4	7.68	79
5–9	7.93	82
10–14	8.03	83
15–65	63.85	656
>65	12.51	128
Total	100	1028

It is necessary to make a number of assumptions about population exposure in order to conduct the QMRA.

7.4.3.1 Evacuation

It was assumed that all residents were at home at the time of the flood and that they all chose to evacuate (rather than remain in their property). It was assumed that residents did not immerse themselves voluntarily in the floodwater (e.g. for recreation or exploration) other than for the purposes of evacuation. Table 7.6 outlines the estimated likelihood of immersion by age, assuming that the flood depth reaches at least 0.65m. These proportions were assumed on the basis that children aged four and under would probably be carried to safety, and that the likelihood of immersion for older children then declines with age due to increasing strength, stability and common sense. Adults between the ages of 18 and 65 were assumed to succumb to the least number of immersions. People aged 65 and over are likely to decline in mobility and stability in flood through the physical effects of aging, hence an immersion rate of 5% was assumed, although they may be more inclined to request assistance from the emergency services. It was assumed that an immersion resulted in a single gulp of water being swallowed (i.e. approximately 30ml for adults and 20ml for children).

Table 7.6. Assumed likelihood of immersion by age

Age group	Estimated immersion (% population)
0–4	0
5–10	10
11–18	3
Adult <65	1
Adult 65 and over	5

7.4.3.2 Clean-up

It was assumed that the majority of residents will wait until the water has dropped considerably, before returning to the flooded property or beginning the remediation process and will continue to live elsewhere until the property is restored (Tapsell pers. comm.). The duration of the clean-up process will depend upon the size of the property, the number of people involved and the extent of the flood damage. The initial clean-up process (when people may be exposed to pathogens) has been assumed to take between one and four days, during this period daily floodwater contact is estimated to be the length of the working day, with those involved returning to alternative accommodation at the end of the day.

The volume of water ingested accidentally through mouthing behaviour is difficult to quantify. Westrell *et al.* (2004) assumed an ingestion of 1 ml in children playing nearby near a source of reused wastewater, and transferring water from hand to mouth. Tanaka *et al.*, 1998 (using data from Asano *et al.*, 1992) assumed that a golfer exposed to a course irrigated in wastewater will ingest 1 ml in a single exposure. It has been assumed that in a clean-up situation 1 ml per hour would be ingested in both children and adults (while children are likely to ingest a greater volume than adults, it is also likely that they would be exposed for a shorter period therefore, for simplicity, the same figures have been assumed).

7.4.3.3 Flood composition

The relative proportions of water from each of the identified flood components outlined in section 7.2 will be incident specific. To illustrate the QMRA it has been assumed that river water comprises the majority of the flood flow (80%), with 10% from urban surface runoff and 10% foul flow.

The pathogen concentration ranges for the riverine and foul flow components are shown in Table 7.7, the urban non foul component has been assumed to act as a diluting component, containing negligible levels of pathogens in comparison to the other components.

Table 7.7. Pathogen concentration parameters for the principal flood components

	Concentration (/litre)	
	River	Foul flow
Campylobacter	0–3300	10–180,000
Cryptosporidium	0–59	1–96
Virus	1–64*	70–3200

* based on a 50th of the sewage value

7.4.4 Risk characterization

This section brings together the estimates of exposure and dose-response for each of the identified hazards to provide an overall estimation of the risk of illness, as shown in Table 7.8.

Table 7.8. Risk characterization summary

Hazard	Exposure	Estimate	Cases of illness	DALYs
Campylobacteriosis	Evacuation	Min	0.008	2.4×10^{-5}
		Mean	0.312	9.4×10^{-4}
		Max	1.691	5.1×10^{-3}
	Clean up	Min	0.076	2.2×10^{-4}
		Mean	7.529	2.1×10^{-2}
		Max	56.67	1.6×10^{-1}
Cryptosporidiosis	Evacuation	Min	0.002	3.9×10^{-6}
		Mean	0.016	3.1×10^{-5}
		Max	0.086	1.4×10^{-4}
	Clean up	Min	0.008	1.1×10^{-5}
		Mean	0.387	5.3×10^{-4}
		Max	3.002	4.1×10^{-3}
Viral enteritis	Evacuation	Min	0.326	4.1×10^{-4}
		Mean	1.578	2.0×10^{-3}
		Max	3.634	4.6×10^{-3}
	Clean up	Min	0.141	1.8×10^{-4}
		Mean	38.16	4.8×10^{-2}
		Max	139.9	1.7×10^{-1}
Overall illness		**Mean**	**48**	$\mathbf{7.2 \times 10^{-2}}$

It can be seen from Table 7.8 that the risk is dominated by viral enteritis contracted during the clean up process, followed by campylobacteriosis contracted during clean up.

7.5 DISCUSSION

In this exploratory QMRA, two flood-related scenarios were examined, namely the risk of illness while evacuating a flooded property (based on the assumed likelihood of immersion) and the risk of illness contracted during the clean up process. As data on pathogen concentrations, floodwater composition and behavioural responses to flooding are limited and likely to be highly variable (and largely event specific) it was necessary to make a number of assumptions. Where possible these were based on an assessment of the literature or limited empirical data acquisition.

The risk of illness was dominated by clean up process, with 46 of the estimated 48 cases of illness being attributable to this scenario. This probably reflects the greater duration of exposure to contaminated water and the assumption (for purposes of the calculation) that doses received over the whole clean up period were cumulative. The dominance of viral enteritis reflects the low infectious dose.

In every case, except one, the estimates were most sensitive (data not shown) to the concentration of pathogens contained within the river water component of the floodwater. The exception being viral infection during the clean up process, where the estimate was most sensitive to the assumed duration of the clean up (i.e. between one and four days).

Applying the literature relative risk value of 2.36 from Chapter 6 (Table 6.2) to this population results in a estimate of 27 cases of illness, which is similar to the overall mean estimate (48) derived by the QMRA method. Although the estimated number of cases of illness is relatively high, it is likely that the majority would not be picked up by the routine surveillance system and an elevation in cases of enteric illness due to flooding may not be identified. Many people would not visit their GP and the GP may choose not to take a faecal sample from those reporting, particularly if the worst symptoms have passed. Finally, where a sample is taken and analyzed, the clinical laboratory might not isolate the pathogen causing the reported symptoms. Unless all of these stages are completed the illness will not be entered onto the surveillance system records in the UK. When added to the population dispersion due to flooding, resulting from those affected moving out of the area to stay in temporary accommodation or with friends and family, caution is prudent in the interpretation of the partial surveillance data available.

The overall estimated number of DALYs (mean estimate) of 7.2×10^{-2}, equates to a score of 7.05×10^{-5} per person, which is just above the screening level of risk (5×10^{-5}) outlined in Chapter 2 (section 2.7). This score, however, assumes that children under the age of 5 will not be exposed to floodwaters as they are carried during the evacuation and they will not be present in the home during the clean up process. The DALY score would be much higher if this group were exposed as they are particularly vulnerable to rotaviral infection.

7.6 REFERENCES

Abt, S.R., Wittler, R.J., Taylor, A. and Love, D.J. (1989) Human stability in a high hazard flood zone. *Water Resources Bulletin* **25**(4), 881–890.

Adak, G.K., Long, S.M. and O'Brien, S.J. (2002) Trends in indigenous foodborne disease and death, England and Wales: 1992 to 2000. *Gut* **51**, 832–841.

Albrechtsen, H.J. (2002) Microbiological investigation of rainwater and greywater collected for toilet flushing. *Water Science and Technology* **46**(6–7), 311–316.

Anon (2007) Killer bugs in flood water. Mirror 26/07/2007. http://www.mirror.co.uk/news/topstories/2007/07/26/killer-bugs-in-flood-waters-89520–19521347/

Arimi, S.M., Fricker, C.R. and Park, R.W.A. (1988) Occurrence of 'thermophilic' campylobacters in sewage and their removal by treatment processes. *Epidemiology and Infection* **101**, 279–286.

Arnone, R.D. and Walling, J.P. (2006) Evaluating *Cryptosporidium* and *Giardia* concentrations in combined sewer overflow. *Journal of Water and Health* **4**(2), 157–165.

Asano, T., Leong, L.Y.C., Rigby, M.G. and Sakaji, R.H. (1992) Evaluation of the California wastewater reclamation criteria using enteric virus monitoring data. *Water Science and Technology* **26**(7–8), 1513–1524.

Bates, P. and Phillips, C.A. (2005) Agricultural practices as a source of *Campylobacter* spp. in river water. *Journal of Environmental Health Research* **4**(1), 17–23.

Birks, R., Colbourne, J., Hills, S. and Hobson, R. (2004) Microbiological water quality in a large in-building, water recycling facility. *Water Science and Technology* **50**(2), 165–172.

Bolton, F.J., Coates, D., Hutchinson, D.N. and Godfree, A.F. (1987) A study of thermophilic campylobacters in a river system. *Journal of Applied Bacteriology* **62**, 167–176.

CEH Centre for Ecology and Hydrology (2007) *The July 2007 floods in England and Wales – A preliminary appraisal issued by the Centre for Ecology & Hydrology*. http://www.ceh.ac.uk/news/press/BriefingnoteJuly2007Floods.html (Accessed 5/9/07).

Chapron, C.D., Ballester, N.A., Fontaine, J.H., Frades, C.N. and Margolin, A.B. (2000) Detection of astroviruses, enteroviruses and adenovirus types 40 and 41 in surface waters collected and evaluated by the information collection rule and an integrated cell culture-nested PCR procedure. *Applied and Environmental Microbiology* **66**(6), 2520–2525.

Crowther, J., Wyer, M.D., Bradford, M., Kay, D. and Francis, C.A. (2003) Modelling faecal indicator concentrations in large rural catchments using land use and topographic data. *Journal of Applied Microbiology* **94**, 962–973.

DEFRA/EA (2006) Flood Risks To People, Phase 2. FD2321/TR1: The Flood Risks to People Methodology (Flood and Coastal Defence R&D Programme).

Drabek, T.E. (2000) The Social Factors that Constrain Human Responses to Flood Warnings. In *Floods, Volume 1* (ed D.J Parker), pp. 361–376, Routledge, London.

Ferguson, C. (2005) Deterministic model of microbial sources, fate and transport: a quantitative tool for pathogen catchment budgeting. PhD University of New South Wales, Australia.

Ferguson, C., de Roda Husman, A.M., Altavilla, N. and Deere, D. (2003) Fate and transport of surface water pathogens in watersheds. *Critical Reviews in Environmental Science and Technology* **33**, 299–361.

Fewtrell, L. and Kay, D. (2007) Microbial quality of rainwater supplies in developed countries: a review. *Urban Water Journal* **4**(4), 253–260.

FSA Food Standards Agency (2000) A Report of the Study of Infectious Intestinal Disease in England. The Stationary Office, London.

GAD (2007) Interim Life Tables. Government Actuary's Department http://www.gad.gov.uk/Demography_Data/Life_Tables/Interim_life_tables.asp accessed February 2007.

Gannon, J.J. and Busse, M.K (1989) *E. coli* and enterococci levels in urban stormwater, river water and chlorinated treatment plant effluent. *Water Research* **23**(9), 1167–1176.

Gibson, C., Stadterman, K., States, S. and Sykora, J. (1998) Combined sewer overflows: a source of *Cryptosporidium* and *Giardia*? *Water Science and Technology* **38**(12), 67–72.

Gruntfest, E. and Ripps, A. (2000) Flash Floods. In *Floods, Volume 1* (ed D.J Parker), pp. 376–390, Routledge, London.

Haas, C. and Eisenberg, J.N.S. (2001) Risk assessment. In: *Water Quality: Guidelines, Standards and Health* (eds L. Fewtrell and J. Bartram), pp. 161–183, World Health Organization Series. IWA Publishing, London.

Haas, C.N., Rose, J.B., Gerba, C. and Regli, S. (1993) Risk assessment of virus in drinking water. *Risk Analysis* **13**, 545–552.

Havelaar, A.H., de Wit, M.A.S., van Koningsveld, R. and van Kempen, E. (2000a) Health burden in the Netherlands due to infection with thermophilic *Campylobacter* spp. *Epidemiology and Infection* **125**, 505–522.

Havelaar, A.H., Hollander, de A.E.M., Teunis, P.F.M., Evers, E.G., Kranen, van H.J., Versteegh, F.M., Koten, van J.E.M. and Slob, W. (2000b) Balancing the risks and benefits of drinking-water disinfection: disability-adjusted life years on the scale. *Environmental Health Perspectives* **108**, 315–321.

Holländer, R., Bullermann, M., Groß, C., Hartung, H., Künig, K., Lücke, F-K. and Nolde, E. (1996) [Microbiological and hygienic aspects of the use of rainwater as process water for toilet flushing, garden irrigation and laundering.] *Gesundheitswesen* **58**, 288–293. [Article in German].

HPA (2007a) Epidemiological Data – Cryptosporidium. http://www.hpa.org.uk/infections/topics_az/crypto/data.htm (accessed 01/11/07)

HPA (2007b) Epidemiological Data – Rotavirus. http://www.hpa.org.uk/infections/topics_az/rotavirus/data.htm (accessed 01/11/07)

Jones, K., Betaieb, M. and Telford, D.R. (1990) Thermophilic campylobacters in surface waters around Lancaster, UK: negative correlation with campylobacter infections in the community. *Journal of Applied Bacteriology* **69**, 758–764.

Jones, K. (2001) Campylobacters in water, sewage and the environment. *Journal of Applied Microbiology* **90**, 68S–79S.

Karvonen, T., Hepojoki, A., Huhta, H-K. and Louhio, A. (2000) The Use of Physical Models in Dam-Break Analysis. RESCDAM Final Report, Helsinki University of Technology, Helsinki, Finland. [Cited in Lind *et al.* (2004)]

Kay, D., Edwards, A.C., McDonald, A.T., Stapleton, C.M., Wyer, M. and Crowther, J. (2007) Catchment microbial dynamics: the emergence of a research agenda. *Progress in Physical Geography* **31**(1), 59–76.

Kay, D., Stapleton, C., Crowther, J., Wyer, M., Fewtrell, L., Edwards, A., McDonald, A.T., Watkins, J., Francis, C.A. and Wilkinson, J. (2008) Faecal indicator organism compliance parameter concentrations in sewage and treated effluents. *Water Research* **42**(1–2), 442–454.

Krikelis, V., Spyrou, N., Markoulatos, P. and Serie, C. (1985) Seasonal distribution of enteroviruses and adenoviruses in domestic sewage. *Canadian Journal of Microbiology* **31**, 24–25.

Lind, N., Hartford, D. and Assaf, H. (2004) Hydrodynamic models of human instability in a flood. *Journal of the American Water Resources Association* **40**(1), 89–96.

McCuin, R.M. and Clancy, J.L. (2006) Occurrence of *Cryptosporidium* oocysts in US wastewaters. *Journal of Water and Health* **4**(4), 437–452.

Mead, P.S., Slutsker, L., Dietz, V., McCaig, L.F., Bresee, J.S., Shapiro, C., Griffen, P.M. and Tauxe, R.V. (1999) Food-related illness and death in the United States. *Emerging Infectious Diseases* **5**(5), 607–625.

Medema, G.J., Teunis, P.F.M., Havelaar, A.H. and Haas, C.N. (1996) Assessment of the dose-response relationship of *Campylobacter jejuni. International Journal of Food Microbiology* **30**, 101–111.

Montemayor, M., Valero, F., Jofre, J. and Lucena, F. (2005) Occurrence of *Cryptosporidium* spp. oocysts in raw and treated sewage and river water in north-eastern Spain. *Journal of Applied Microbiology* **99**, 1455–1462.

Moore, J.E., Caldwell, P.S., Millar, B.C. and Murphy, P.G. (2001) Occurrence of *Campylobacter* spp. in water in Northern Ireland: implications for public health. *The Ulster Medical Journal* **70**(2), 102–107.

Myrmel, M., Berg, E.M.M., Grinde, B. and Rimstad, E. (2006) Enteric viruses in inlet and outlet samples from sewage treatment plants. *Journal of Water and Health* **4**(2), 197–209.

Navin, T.R. and Hardy, A.M. (1987) Cryptosporidiosis in patients with AIDS. *Journal of Infectious Diseases* **155**(1), 150.

Norwich Union (2007) A Guide to Flood Resilient Repairs. Accessed online 22.08.07 http://www.floodresilienthome.com/downloads/NU_FLOODGUIDE_A5.pdf

NRMMC (2006) Australian Guidelines for Water Recycling: Managing Health and Environmental Risks (Phase 1). Natural Resource Management Ministerial Council, Environment Protection and Heritage Council, Australian Health Ministers Conference, Canberra, Australia.

Obiri-Danso, K. and Jones, K. (1999) Distribution and seasonality of microbial indicators and thermophilic campylobacters in two freshwater bathing sites on the River Lune in northwest England. *Journal of Applied Microbiology* **87**(6), 822–832.

Olivieri, V.P., Kruse, C.W., Kawata, K. and Smith, J.E. (1977) Microorganisms in Urban Stormwater. USEPA Report EPA-600/2-77-087. US Environmental Protection Agency.

ONS Office for National Statistics (2007) Mortality Statistics: Cause (Series DH2) accessed 01.11.07 http://www.statistics.gov.uk/StatBase/Product.asp?vlnk=618

Parker, D.J. (1991) The Damage-Reducing Effects of Flood Warnings. Report prepared for Halcrow/National Rivers Authority (Anglian Region) Regional Telemetry Scheme Appraisal, Flood Hazard Research Centre, Middlesex University, UK. [Cited in Tapsell and Tunstall (2000)]

Pina, S., Puig, M., Lucena, F., Jofre, J. and Girones, R. (1998) Viral pollution in the environment and in shellfish: human adenovirus detection by PCR as an index of human viruses. *Applied and Environmental Microbiology* **64**(9), 3376–3382.

Pitt, R. and McLean, J. (1986) Toronto Area Watershed Management Strategy; Humber River Pilot Watershed Project. Final Report. Ontario Ministry of the Environment, Toronto.

Rodgers, P., Soulsby, C., Hunter, C. and Petry, J. (2003) Spatial and temporal bacterial quality of a lowland agricultural stream in north east Scotland. *Science of The Total Environment* **314–316**, 289–302.

Rose, J.B. (2007) Water reclamation, reuse and public health. *Water Science and Technology* **55**(1–2), 275–282.

Rosenthal, U. and Bezuyen, M.J. (2000) Flood Emergency Management in Developed Countries. In *Floods, Volume 1* (ed D.J Parker), pp. 340–350, Routledge, London.

Rusin, P., Enriquez, C., Johnson, D. and Gerba, C. (2000) Environmentally Transmitted Pathogens. In *Environmental Microbiology* (eds R.M. Maier, I.L. Pepper and C.P. Gerba), Academic Press, San Diego, CA.

Savill, M.G., Hudson, J.A., Ball, A.A., Klena, J.D., Scholes, P., Whyte, R.J., McCormick, R.E. and Jankovic, D. (2001) Enumeration of *Campylobacter* in New Zealand recreational and drinking waters. *Journal of Applied Microbiology* **91**, 38–46.

Schillinger, J.E. and Gannon, J.J. (1985) Bacterial adsorption and suspended particles in urban stormwater. *Journal of the Water Pollution Control Federation* **57**(5), 384–389.

Selvakumar, A. and Borst, M. (2006) Variation of microcosm concentrations in urban stormwater runoff with land use and seasons. *Journal of Water and Health* **4**(1), 109–124.

Simmons, G., Hope, V., Lewis, G., Whitmore, J. and Gao, W. (2001) Contamination of potable roof-collected rainwater in Auckland, New Zealand. *Water Research* **35**, 1518–1524.

Smith, K., Fewtrell, L. and Kay, D. (2007) Examination of the microbiology of urban flood water components. FRMRC RPA 6.2 working paper. University of Wales, Aberystwyth.

Stapleton, C.M., Wyer, M.D., Crowther, J., McDonald, A.T., Kay, D., Greaves, J., Wither, A., Watkins, J., Francis, C., Humphrey, N. and Bradford, M. (2008) Quantitative catchment profiling to apportion faecal indicator organism budgets for the Ribble system, the UK's sentinel drainage basin for Water Framework Directive research. *Journal of Environmental Management* (in press).

States, S., Stadterman, S., Ammon, L., Vogel, P., Baldizar, J., Wright, D., Conley, L. and Sykora, J. (1997) Protozoa in river water: sources, occurrence and treatment. *Journal of the American Water Works Association* **89**(9), 74–82.

Tani, N., Dohi, Y., Kurumatani, N. and Yonemasu, K. (1995) Seasonal distribution of adenoviruses, enteroviruses and reoviruses in urban river water. *Microbiology and Immunology* **39**(8), 577–580.

Tanaka, H., Asano, T., Schroeder, E.D. and Tchobanoglous, G. (1998) Estimating the safety of wastewater reclamation and reuse using enteric virus monitoring data. *Water Environment Research* **70**(1), 39–51.

Tapsell, S.M. and Tunstall, S.M. (2000) The Health Effects of Floods. In *Floods, Volume 1* (ed D.J. Parker), pp. 172–187, Routledge, London.

Teunis, P.F.M., Heijden, van der O.G., Giessen, van der J.W.B. and Havelaar, A.H. (1996) The dose-response relation in human volunteers for gastro-intestinal pathogens. RIVM Report 284550002, Bilthoven, Netherlands.

Tulve, N.S., Suggs, J.C., McCurdy, T. and Moya, J. (2002) frequency of mouthing behaviour in young children. *Journal of Exposure Analysis and Environmental Epidemiology* **12**, 259–264.

van Duin, M.J. and Bezuyen, M.J. (2000) Flood Evacuation During the Floods of 1995 in The Netherlands. In *Floods, Volume 1* (ed D.J. Parker), pp. 351–360, Routledge, London.

VGDHS (1999) Victoria Burden of Disease Study. Victorian Government Department of Human Services, Australia.

Westrell, T., Schönning, C., Stenström, T.A. and Ashbolt, N.J. (2004) QMRA (quantitative microbial risk assessment) and HACCP (hazard analysis and critical control points) for management of pathogens in wastewater and sewage sludge treatment and reuse. *Water Science and Technology* **50**(2), 23–30.

WHO (2004) Guidelines for Drinking-water Quality. Third Edition. Volume 1: Recommendations. World Health Organization, Geneva, Switzerland.

Wilkinson, J., Kay, D., Wyer, M.D. and Jenkins, A. (2006) Processes driving the episodic flux of faecal indicator organisms in streams impacting on recreational and shellfish harvesting waters. *Water Research* **40**(1), 153–161.

Wyer, M.D., Crowther, J., Kay, D., Francis, C.A., Watkins, J. and Bradford, M. (2000) Bathing water compliance at north Ceredigion beaches: statistical modelling and catchment investigations. Report of the Interreg Project. 44pp (plus figures and tables). CREH, University of Wales, Aberystwyth.

Wyer, M.D., Stapleton, C., Kay, D., Francis, C., Watkins, J. and Rushby, L. (2001) *Cryptosporidium*, *Giardia* and faecal indicator organisms in surface water and sewage effluent from the Irvine catchment, Ayrshire. Report to the DWI, West of Scotland Water, SEPA, South Ayrshire and North Ayrshire councils. CREH, University of Wales, Aberystwyth.

8

Mupfure irrigation project, Zimbabwe. HIA of a water resources development

Flemming Konradsen, Moses Chimbari and Peter Furu

8.1 BACKGROUND

Since King Menes dammed the Nile some 5000 years ago to improve the agricultural production, millions of hectares of land have been put under irrigation. At present, it is estimated that worldwide about 270 million hectares are irrigated (Molden 2007). Irrigation has not only enabled agricultural production in arid and semi-arid areas but it has also increased crop yield and extended growing seasons on lands already under cultivation.

The linkage between social and environmental changes and the development of water resources and the possible adverse health effects on surrounding communities have been debated for decades (e.g. Lanoix 1958). The most pronounced negative health consequences tend to be those related to vector-borne disease. The actual health impact of a specific project is determined by a variety of interacting factors, among which geographical location, technical design, human migration, choice of crops, operation and maintenance practices and pre-project health and socio-economic status are considered to be of importance. This chapter outlines an early example of an HIA, which was conducted in Zimbabwe in 1993 based on material written at the time.

8.2 PROJECT DETAILS

The HIA was conducted on the proposed Mupfure Irrigation Project (MIP), a small-scale irrigation project to be implemented in the Mashonaland Central Province of Zimbabwe, located approximately 60 km north of Bindura, the provincial capital. The HIA is written (as it was at the time) in the future tense.

The project area is between 1000 and 1080 meters above sea level (Highveld region) with undulating rocky hills to the south and east. Prior to the project proposal, most of the area had already been cleared of vegetation and was being used for rainfed crop production.

The project is expected to be in operation for approximately 40 years before rehabilitation, with the construction phase lasting for three years. On completion, the MIP is to be a farmer-managed project, with the farmers being responsible for the day-to-day management of the scheme (assisted by government officials). The feasibility study concluded that the limitations of rainfed cropping, largely due to too little rainfall, are the major constraints hampering further improvement of the living standard of the rural population in the area. Therefore, the overall conceptual approach of the project is to utilize irrigation as an economic means of improving the general standard of living of the local inhabitants, with intensified land-use and increased crop productivity.

It is proposed that an earth fill dam (Goora Dam) will be constructed on Zvisokwe river, a tributary of Mupfure river, to supply water for MIP. The dam wall is to be 23 m high, with an estimated storage capacity of the reservoir of 4.3 million m^3, covering up to 100 ha of land. Water from the reservoir will be pumped into a conveyance canal and, by gravity, to a night storage pond. The water will then be distributed to secondary and tertiary canals through a main canal. Siphons will be used to deliver water from the tertiary canal into irrigation furrows on the fields. It is planned that all canals, except the tertiary canal, will be concrete lined. The proposed drainage system uses existing natural streams.

When fully developed, it is expected that the project annual water demand for irrigation and other use (e.g. fish ponds) will amount to 1.9 million m^3, with a focus on high value cash crops, such as tobacco, cotton, groundnuts and beans, with a smaller part of the land allocated for subsistence crops. The project is designed to result in minimum interruption to the local population and it is estimated that only four families will have to be resettled as a result of the construction of the Goora dam.

8.3 SCREENING

Before the decision was made to prepare a full-scale HIA, the project was subjected to a screening process to determine possible health hazards and their implications and to determine the merit of conducting a full HIA. The screening process identified a number of health hazards and health opportunities that were subsequently included in the terms of reference for the full-scale HIA. This step was based on readily available information, such as project documents, general country information, national statistical yearbooks, reference handbooks, maps and information regarding health hazards typically associated with this type of project (Konradsen *et al.* 1997).

8.4 SCOPING

Scoping was initiated following the screening process, in order to decide which health concerns, health hazards and health benefits should be further assessed. Scoping also defined the timing and boundaries of the HIA and sources of expertise required. Further data and expert opinion were gathered from nongovernmental organizations, scientific publications as well as national and international agencies. The result of the scoping process was a set of terms of reference for the full-scale HIA.

8.5 FULL-SCALE HIA

The health risk associated with each of the identified hazards was assessed according to the following:

- environmental factors – such as those associated with the project construction and use;
- vulnerable communities – groups exposed to the potential hazard, their likely susceptibility and behaviour and demographic changes resulting from the project that could affect vulnerability; and

- capabilities of the health protection agencies – their ability to detect, monitor, inform, safeguard and mitigate health risks.

The information needed to carry out the HIA was extracted from existing project documents, scientific papers and through the use of rapid assessment procedures where key informants were identified at central, provincial, district and village levels. A series of interviews (48 individuals) and focus group discussions were conducted (on five occasions) at the different levels. No structured questionnaires were used but an outline of questions that would solicit information required were prepared beforehand. Another important part of the rapid assessment procedure consisted of site visits and general observations in and around the project area. Relevant information was also gathered from internal documents and statistics provided by the Ministry of Health and Child Welfare, the Irrigation Department, the Water Department, the national Economic Planning Commission, bilateral donor organizations, the University of Zimbabwe and the World Bank. The field studies were carried out during February, March and April 1993.

The following sections describe the main findings of the HIA focusing on important health concerns of the local communities.

8.5.1 Schistosomiasis

8.5.1.1 Environmental factors

A number of environmental factors were identified which are considered to increase the likely transmission of schistosomiasis as a result of the MIP.

- Snail population: surveys conducted prior to, and as part of, the HIA consistently found the snails *Bulinus globosus* and *Biomphalaria pfeifferi*, the respective intermediate hosts of *Schistosoma haematobium* and *S. mansoni*, in the project area.
- Flood reduction: prior to the project, the snails found in the Zvisokwe river were subject to seasonal floods which washed them downstream. The construction of Goora dam will stop the seasonal flooding and, as a result, is likely to increase the number of snails in the small pools created in the river bed.
- Canal water velocity: the design of the main and secondary canals should (due to the water velocity) discourage snail breeding. If canals are not regularly cleared of obstructions, like silt and plant material, the water velocity is likely to drop and thus promote snail breeding. Silt and

vegetation will not only decrease water velocity but may also provide snails with shelter and food.

- Dam creation: the colonization of the Goora Dam by snails is expected to be rapid as the snails are already present in the river.
- Other impoundments: the night storage pond is expected to eventually be colonized by snails, as has been the case in other similar ponds in the country (Chandiwana *et al.* 1998).
- Drainage canals: the use of natural waterways as the drainage system for the irrigation system will turn seasonal waterways into perennial streams thus providing a favourable habitat for snails. Furthermore, it can be expected that excess fertilizer from the fields will end up in the drainage canals and create increased plant growth, making the habitat more favourable for snails.
- Irrigation practices: the system of land allocation is expected to cause water management problems as all irrigation blocks will require water during the same periods, meaning that it will not be possible to dry canals regularly in order to reduce snail population densities.

8.5.1.2 Community vulnerability

Parasitological surveys conducted at the local primary school showed an often high prevalence of schistosomiasis, especially *S. haematobium*. This demonstrates that the general population is vulnerable to the disease and that any changes in the environment, which favour snail breeding, could lead to further exposure resulting in a further increased prevalence and intensity of disease.

Borehole water in the area is hard and it has been noted that people prefer to use river water for washing and bathing because the hard water from the borehole does not foam easily and, therefore, requires larger amounts of soap. An increase in the amount of surface water due to the irrigation project, in a situation where existing safe water supplies are of low popularity and inadequate, will promote human contact with unsafe water. In addition, the sanitation in the area is very poor and environmental contamination of water bodies with schistosome eggs from infected people, who have no access to latrines, and human contact with infected water are central factors in transmission. Although the project includes plans to build additional latrines in the fields, the numbers are considered to be inadequate. Although good latrine coverage will reduce the number of schistosome eggs reaching the water, it will only have a minor effect on the transmission of *S. haematobium* as observations, made elsewhere in the country (Chimbari *et al.* 1992), indicate that the majority of the population will not go to the latrines just to urinate.

Irrigation plot holders will be vulnerable to infection because of their irrigation-related activities. This vulnerable group will be mainly adults who will spend most of their time on risky activities such as watering, canal cleaning and general maintenance of the irrigation infrastructure. This group will, as a result of the development of the MIP, be exposed to schistosomes for the greater part of the year (except from the end of May to August, when the temperatures are too low for the development of the larval stages of the parasite and snail numbers are generally low).

Vulnerability among children is likely to increase significantly because of the development of the MIP. The children, mostly boys, will be exposed to infections during swimming and fishing, while the girls will have increased water contact while carrying out domestic activities.

Members of the Apostolic Faith church constitute approximately one third of the total population that may be affected by the project (or an estimated 1100 people). Members of this church associate water with holiness and they are frequently in contact with natural water sources because of their religious rituals, which are performed using such waters. They form a very vulnerable group because of their belief that diseases, including schistosomiasis, can be cured by praying for the infected.

The cropping pattern in Mupfure will change from a summer rainfall dependent system to a perennial system based on availability of irrigation water. The workload of the farmers will increase and, in order to cope during labour-intensive periods (weeding and harvesting periods), they will seek assistance from relatives or casual labourers from outside the project area. Except for the month of June, the labour-intensive period coincides with the period of possible schistosomiasis transmission in the study area; resulting in an increased risk of contracting the disease.

A significant number of the members of the Apostolic Faith group resent any form of medical treatment and therefore remain an important reservoir of infection, contributing significantly to continued contamination of the environment with parasite eggs. Furthermore, the migrant labourers coming to the area could become an important reservoir of infection if they have not been treated or been part of any health promotional activities in their home areas.

8.5.1.3 Capabilities of the health protection agencies

There is no mechanism for identifying snail breeding sites and no capacity for snail control activities in the project area despite the fact that the snails are present in large numbers. In addition, it is not possible to diagnose *S. mansoni* at the local health centres (due to a lack of microscope facilities) and there are no drugs available locally for the treatment of *S. mansoni*.

8.5.1.4 Schistosomiasis summary

The development of the MIP is likely to lead to environmental changes resulting in an increased intensity of schistosomiasis transmission. However, changes in the transmission season are less likely to take place as temperatures in winter become sufficiently low to reduce or even completely suppress the development of the larval stages. Increased transmission is especially likely to take place close to human settlements neighbouring irrigation canals and the Goora Dam, and where domestic and recreational water contact are probably intense.

The most important reasons for the continuously high prevalence of schistosomiasis in the project area, identified during the HIA are:

- a lack of facilities for diagnosing *S. mansoni*;
- a lack of drugs for treating *S. mansoni* infections;
- an absence of a snail control programme;
- an ineffective safe water supply and sanitation programme; and
- a lack of a well-coordinated health education and disease monitoring programme.

8.5.2 Malaria

Anopheles arabiensis is the principal vector of malaria transmission in the project area. It breeds well in shallow, open, sun-lit habitats, ranging from borrow-pits, brick-pits, car-tracks and hoof prints, to rice fields and irrigation drainage canals (Le Sueur and Sharp 1988). *An. arabiensis* quickly and readily exploits any temporary pools, especially in the absence of predators. Breeding of the vector mosquito largely depends on the availability of water and temperature. Hence the density of mosquitoes will vary during the year.

8.5.2.1 Environmental factors

Numerous borrow pits are usually created during the construction phase of dams and irrigation schemes. In most cases, construction contractors do not fill in the borrow pits at the end of the construction phase. These then fill with water during the rainy season and become good habitats for the breeding of *An. arabiensis*. If this happens during the development of MIP, then the number of potential breeding sites for the vector mosquito will increase significantly.

The proposed irrigation scheme will create marshy conditions in the drainage system because of continued water requirements. Hence water will always be found in the canals and the drainage system. This will create good habitats for mosquito breeding particularly alongside the drains where small pools of water are likely to be formed due to the relatively low water velocity and extensive seepage. No mosquito breeding is expected to take place in the fields since none of the proposed crops require permanent standing water.

It has been estimated that up to roughly 60% of the Goora Dam shoreline could become suitable for mosquito breeding. Mosquito habitats close to the shoreline may experience an elevated humidity level resulting from local changes in microclimate created by the reservoir. An increased atmospheric humidity may increase the longevity of adult anophelines and thereby increase the vectorial capacity.

In a number of localities in the southern hemisphere, the heavy use of pesticides to control agricultural pests has had the unforeseen effect of producing insecticide-resistance in mosquitoes. Some of the crops proposed for the MIP (e.g. cotton and tobacco) require a variety of agrochemicals, such as organochlorines and organophosphates, for pest control (see section 8.5.3). If used intensively, these agrochemicals may induce cross resistance, in mosquitoes, to insecticides in the same chemical group.

8.5.2.2 Community vulnerability

In general, people repeatedly exposed to malaria in endemic areas develop a degree of immunity. However, in unstable malaria areas of low endemicity where the immunity in the indigenous population is low severe infection may occur in all age groups. The project area is located within an area of seasonal malaria transmission. An active parasite survey carried out in the region estimated malaria prevalence to be about 1%. Consequently, the general population of the project area is vulnerable to malaria infections and their malaria immune status is expected to be low.

Environmental changes associated with the development of the MIP may result in malaria outbreaks due to the increased number of vectors and lack of protective immunity in the human population. The most vulnerable group to outbreaks will be children under the age of five and pregnant women (because of their general weak immune status).

The majority of houses in the project area are brick roundavels (round huts), with grass thatched roofs. The houses have eaves and there are usually small open windows in the walls for ventilation. The windows cannot be closed and are just an opening where one or two bricks have been left out during construction. This design allows movement of mosquitoes in and out of the room, thus, exposing the people sleeping inside to mosquito bites. Furthermore, the grass roof and dark clay walls provide good resting places for mosquitoes.

Construction workers coming to the area will be living in tents and tin huts close to the construction site and will be exposed to more mosquito bites when borrow pits fill with water. Similarly, 'camp followers' (e.g. prostitutes and tradesmen) will become a vulnerable group during this period. During the construction of the Goora Dam, skilled manpower coming from malaria free areas will form a vulnerable group as they may be more susceptible to the malaria parasite. Those coming from malaria endemic areas may increase the parasite reservoir in the project area and hence increase vulnerability of the entire community. Furthermore, an increased migration will increase the likelihood of chloroquine resistant strains of *Plasmodium* being introduced into the project area. As described in section 8.5.1.2, a large number of people can be expected to come to the area to assist relatives in harvesting and cotton picking. The weeding of summer crops will coincide with malaria peak transmission period.

8.5.2.3 Capability of health protection agencies

Surveillance of malaria outbreaks is poor and there are no indications that it will improve in the near future. The surveillance system is based on compilation of information on clinical cases recorded at the health centres. The information, however, is analysed at provincial level and feedback to ward level is very poor, often taking two or three months. Surveillance is also complicated by the fact that blood slides taken from clinical cases cannot be examined at the health centre, due to lack of facilities. Most of the villages in the project area are not close to health centres, and suspected malaria cases may not reach the health centres as chloroquine is normally available in local shops. Also, members of the Apostolic Faith do not believe in treatment and thus do not report cases. As a result, the health delivery system does not receive sufficient information to keep the malaria situation under control.

Although the majority of the community is aware of how malaria is transmitted and how it can be prevented and cured, the preventive measures are often too expensive or not easily available. For example, window screens, bednets and mosquito repellents are not easily accessible for the community. The health education given is, therefore, less effective as the personal protection methods emphasized cannot be applied.

8.5.2.4 Malaria summary

The total effect of the anticipated environmental changes resulting from the MIP will lead to an increased number of malaria-transmitting mosquitoes throughout the year (except for June and July, where temperatures are likely to be too low for extensive breeding).

The villagers living close to the drains and the Goora Dam are likely to be a highly vulnerable group due to their increased exposure to mosquito bites. Furthermore, the extended human migration will increase the likelihood of chloroquine resistant strains of *Plasmodium* being introduced into the project area.

The ability of the local health services to deal with the present malaria problem is hampered by the lack of a good surveillance system to report outbreaks, an inefficient mosquito control programme and the long distance from the project area to health centres. Furthermore, preventive measures for personal protection are often very expensive and not easily available.

8.5.3 Agrochemical poisoning

8.5.3.1 Environmental factors

Agricultural practices in the project area will change considerably as a result of the development of MIP. Shifting from dryland farming to irrigation will reduce the area of land per family. However, production will be intensified and total output is expected to increase markedly. Increased production will depend upon application of agrochemicals, mainly insecticides and fertilizers, such as ammonium nitrate. A large proportion of the chemicals used in the irrigated fields is likely to end up in the waterways surrounding the fields, especially the drainage system.

Maize is grown by all families in the project area, with cotton and groundnuts as the second most common crops. In addition, vegetables are grown in small plots close to homesteads. After the project implementation it is expected that the cropping programme will change to that shown in Table 8.1.

The crops in the rotation are dominated by marketable products. Maize will be used both as staple food for the families and to generate income. At present, chemicals are being used to control pests (like white flies, mites and aphids) in the cotton fields. Chemicals are also applied to tobacco to control cutworm and aphids. For the rest of the crops there is no regular use of pesticides, and no herbicides are applied to any of the crops.

As a result of the shift in production, from mainly food crops to the production of cash crops, plans have been made to apply a large number of different chemicals to ensure good output of high quality cash crops. The chemicals will be used during seedbed preparation and for controlling diseases and pests in the crops. Weed control will be done manually.

Table 8.1. Estimated present and future crop yield per farmer

Crop	Present yield (tons)	Projected irrigated yield (tons)	Total irrigated area (hectares)
Beans	0	1.00	85.0
Cotton	0.40	0.33	21.3
Burley tobacco	<0.03	0.63	42.5
Groundnut	0.27	0.50	21.3
Sunflower	0.08	0	0
Maize	1.80	2.63	63.8
Green mealies	0	2000 cobs	21.3

Chemicals will probably be used most intensively for growing burley tobacco, cotton and beans. Out of the 20 different pesticides that have been proposed for routine application in the irrigation scheme, the potentially most hazardous chemicals will be applied in cotton and tobacco fields; they have been rated as moderately hazardous. Dimethoate will be used to control red spider mite in cotton and tobacco fields. The most likely route of human contamination is absorption through the gastrointestinal tract where it can cause an immediate effect on the central nervous system. Death is usually due to respiratory arrest. However, long-term chronic effects are less likely to result as the chemical is rapidly excreted via the urine. Chlorphyrifos will be used to control false wireworms in tobacco fields. It is of utmost importance that this compound is stored away from foods since it is easily adsorbed through the gastrointestinal tract and can cause immediate poisoning. Endosulfan will be used to combat aphids in cotton fields. The compound is readily adsorbed through the skin, gastrointestinal tract and by inhalation of aerosol.

8.5.3.2 Community vulnerability

The human population obtain domestic water from the drainage canals. These are likely to be contaminated with agrochemicals, if provision for adequate safe water supplies is not made. Although the chemicals ending up in the drains are likely to be diluted, and may not present any danger of acute poisoning, they could have a long-term effect.

The farmers who use the agrochemicals will be exposed to an occupational health hazard. This is not only associated with spraying operations, but is also relevant to personnel engaged in mixing, diluting and transporting the chemicals, as well as the washing of equipment after use. Males usually spray the crops and are, consequently, more exposed to chemical poisoning than females. However, people working in fields adjacent to those being sprayed and the children carried on the mother's back, may be exposed through inhalation of the mist or vapour produced during the spraying process. The majority of farmers will have received only basic training in the safe use and handling of pesticides. However, the most vulnerable group, in relation to the use of agrochemicals, is the casual labourers who usually have received no training in any aspect of chemical handling.

The farmers in the project area do not have safe storage facilities to keep their stock of chemicals, so they are kept around the home. This will become a health risk at the household level. An increased stock of different chemicals in their homes will inevitably lead to an increased number of cases of accidental poisoning, especially among children and young people. At the surrounding commercial farms, it is common to see empty chemical containers being used for purposes such as fetching water and storing food items. This habit may well spread to the project area and become a potential health problem. In addition, families unaware of the hazard of chemicals could also be poisoned if vegetables and other foods are not cleaned properly after they have been sprayed with pesticides.

8.5.3.3 Capability of health protection agencies

The current use of protective clothing among farmers in the area during spraying is very limited. A few farmers use rubber boots and thick plastic trousers when spraying their cotton fields but this is, apparently, not meant to be a safety measure but an all purpose clothing and does not reflect a general awareness. The high price of protective clothing makes it difficult for its use to be promoted.

Agricultural extension workers[1] include safe handling of pesticides in their training programme for farmers, but safe storage and disposal are not included. The agricultural extension workers themselves have a limited theoretical knowledge of the possible health consequences of the use of agrochemicals, with most of their knowledge coming from practical work experience.

While producers of pesticides are required, by law, to label their products, the low adult literacy rates mean that many of the farmers will not be able to read the label on the container.

There is no safe disposal system to handle excess chemicals or used containers within the project area and there are no plans for establishing such a system as part of the project. There is a risk of left-over containers and unused chemicals accumulating within the project area.

Since the use of pesticides is rather limited in the project area, the staff at the health centres have little experience in diagnosing and treating patients suffering

---

[1] Field-based staff, providing outreach services.

from agrochemical poisoning. Cases are usually only diagnosed if the farmers themselves can see the relation between their illness and the use of a certain chemical (or in cases where the chemicals have been used in connection with suicide attempts). It is known from other southern African countries that pesticide poisoning often goes undiagnosed because some of the symptoms, vomiting, abdominal cramps and chest pains, are readily associated with other common ailments. In addition, because pesticides can attack the central nervous system, some of the major symptoms (e.g. restlessness and irrational behaviour) can lead to misdiagnosis as mental illness.

Although cases of pesticide poisoning, among workers from the nearby commercial farms, have been observed at one of the health centres, these cases are simply categorized under 'all other diseases'. This lack of an adequate reporting and monitoring system makes it difficult to identify the most vulnerable segments of the population in order to organize specifically targeted health education.

8.5.3.4 Agrochemical poisoning summary

The implementation of the project will lead to increased use of agrochemicals and a large number of new groups of chemicals will be introduced as a result of the increased focus on cash crops and the intensified production. The most vulnerable group is likely to be the farmers carrying out the spraying. Furthermore, the increased use of agrochemicals will inevitably lead to an increase in the number of cases of accidental poisoning due to a lack of storage facilities.

The capacity of the established health system to deal with this increased risk of poisoning is rather limited. This is mainly due to the fact that local health staff lack experience and training in dealing with chemicals as potential health hazards, both in relation to treatment and health education. Furthermore, the absence of systematic data collection makes it difficult to establish an effective monitoring programme to identify vulnerable groups.

8.5.4 Sexually transmitted diseases (STDs)

8.5.4.1 Environmental factors and community vulnerability

It is estimated that some 40 to 50 engineers and specialized workers will be brought into the project area to construct the Goora dam wall, the reservoir and to do minor road works. Furthermore, between 15 and 20 professionals and specialists will be needed for the construction of the irrigation scheme and 150 to 200 local workers will be recruited as casual labourers. A large percentage of the local workers will be recruited from the immediate surrounding villages or at least from neighbouring areas. To accommodate construction workers, from outside the area, temporary settlements will be established. The construction workers may be considered a high risk group, with a higher prevalence of STDs (including HIV), than the population in general, as they often live separated from their partners and travel extensively. Furthermore, the relatively high cash income of the construction workers makes it possible for them to pay prostitutes.

Shortly after the establishment of temporary settlements in connection with major infrastructure or water resource development projects it is common to see camp followers settle down in and around the settlements. The camp followers often consist of prostitutes and tradesmen hoping to make a living by servicing the construction workers. Prostitutes are at high risk of contracting STDs and the disease prevalence within this group is higher than among most other population groups.

There is a risk that the influx of people with a high prevalence of STDs to the project area could lead to an increased number of cases among the local population, since these groups will not live in isolation.

8.5.4.2 Capacity of the health protection agencies

Before treating STDs the health personnel usually require the patient to bring their partner to be examined as well, but this is not always feasible. Most of the health education is done on a personal basis between the health personnel and the patient when treatment is given. More widespread health educational work to promote safe sex seems to be relatively limited or virtually non-existent.

The most common STDs (gonorrhoea, syphilis) are usually diagnosed at the health centres based on clinical symptoms and patients are treated accordingly. In order to diagnose HIV, blood samples have to be forwarded for further analysis at central level but this is not done on a routine basis and most cases of HIV positives are not likely to be identified at an early stage. A large proportion of the STD-affected population seek treatment from traditional healers, which means that it is difficult for the established health system to monitor the number of cases in the area. Thus, it is difficult to plan an efficient health education directed towards the most vulnerable groups.

8.5.4.3 STDs summary

The construction workers and prostitutes are likely to be highly vulnerable groups. However, in the long term, the entire population may be at risk. The capability of the local health services to monitor the spread of STDs in the catchment area is rather limited, especially in the case of HIV where no diagnostic facility is immediately available. Furthermore, the resources available at the local health centres make it difficult for the staff to carry out a well coordinated and efficient health promotional campaign aimed at protecting the local population and the temporary settlers from the increased risk of contracting STDs.

8.5.5 Water-washed diseases

The construction of irrigation canals close to human settlements and the creation of the dam will increase the water supply and make it more reliable and easily accessible throughout the year. An increased amount of accessible water is likely to lead to improved hygiene among the communities living close to the canals and the Goora Dam. The number of hygiene-related skin and eye diseases may therefore decrease as a consequence of the project. However, the number of enteric diseases will continue to be a significant problem due to the lack of a safe drinking water supply and the risk of epidemics may increase if the population use contaminated canal water for drinking.

8.5.6 Nutritional aspects

It is concluded that the planned production of maize under irrigation will be sufficient to satisfy the staple food requirements for even the larger families among the project beneficiaries. Furthermore, the total production of vegetables and meat is expected to remain unchanged. The increased household income will make it possible for the project beneficiaries to allocate more money for food and improve the diet for their families.

Generally, the project will lead to improved food security and limit the consequences of dry spells and pests on agricultural production. However, if the project turns out to be less economically viable than expected there could be a tendency to increase the amount of cash crops and thus change the nutritional impact. During the construction phase food production could decrease as a result of work on project structures in the fields. During this period it might be necessary for the future project beneficiaries to import food. The project may have a negative effect on groups of project non-beneficiaries if the future land use plan deprives this group of arable land.

8.5.7 Other health impacts

During the screening process (section 8.3) psychosomatic diseases and injuries were also examined. However, neither of these was expected to have a large impact and so they were not considered in the full HIA.

8.6 HEALTH IMPACT ASSESSMENT SUMMARY

Based on the health impact assessment, a summary of the expected health implications of implementing MIP is shown in Table 8.2.

8.7 SAFEGUARDS AND MITIGATION MEASURES

Based upon the findings of the HIA and follow-up discussions, a number of safeguards and mitigation measures were identified; these are summarized in Table 8.3 (adapted from Konradsen *et al.* 1997).

8.8 HIA CONCLUSIONS

The project showed that:

- there were adequate data available for a sound HIA;
- integration of HIA into project feasibility studies is prudent; and
- while complexity and methodological problems exist, a systematic evaluation of health impacts yields valuable results.

8.9 MUPFURE REVISITED

It is important to emphasise that the HIA presented here has to be seen as a historical example, using an approach valid at the time of its completion. The terminology and methodology of HIA has changed considerably since the completion of the MIP HIA. For example, in the HIA presented in this chapter we focus on the specific diseases (e.g. schistosomiasis) as the hazard but in the more current HIAs the emphasis is on the agent or cause of the disease as a hazard - and the disease itself as the resulting health outcome. Also, in more recent approaches to HIAs, risk factors will be more central to the analysis. In this way, the discussion of environmental factors will specifically focus on

Table 8.2. Health impact summary

Health hazard	Environmental factors	Vulnerability	Capability of protective agencies	Health Risk
Schistosomiasis	Increased number of snail habitats close to domestic areas and in the working environment	Children swimming in canals, women doing laundry and fetching water, fishermen and farmers working in the scheme	Curative: moderate, drugs only available for *S. haematobium* Surveillance: moderate, no regular surveys and no microscopist Vector control: none	Increased risk – especially in the case of *S. mansoni*
Malaria	Increased number of mosquito breeding habitats close to human settlements. Approximately 2 month extension of the transmission season.	Susceptibility is high for all groups, but under fives and pregnant women are especially vulnerable. Bednets are not used and houses are not being mosquito proofed.	Curative: chloroquine available at health centres (and local shops) but no second line drug Surveillance: poor, long delays in processing information and no monitoring of drug resistance Vector control: moderate, some provision for chemical control in case of epidemics but no control by environmental management	Increased risk
Agrochemical poisoning	Intensified production and increase in cash crops will lead to increased use of chemicals	High occupational risk for irrigation farmers, increased risk of accidental poisoning of children around the households. Generally poor educational level regarding the use of chemicals.	Curative: very limited Preventative: limited, poor storage and disposal facilities, some training of farmers	Increased risk

Table 8.2. *Continued*

Health hazard	Environmental factors	Vulnerability	Capability of protective agencies	Health Risk
Sexually transmitted diseases (STDs)	Increased immigration during construction period and change in social environment	High among construction workers and prostitutes, but may spread to the surrounding community	Curative: adequate for most common STDs, but none for HIV Preventative: poor Surveillance: poor	Increased risk, especially during the construction period
Malnutrition	Increased income and agricultural output, leading to greater food security for project beneficiaries	High among under fives and landless. Reduced vulnerability with increasing income but unequal entitlement to cash earnings expected.	Curative: limited, but active child supplementary feeding programmes Preventative: moderate, mother and child programme and active education and awareness programmes	Reduced risk, possibility for improved nutritional status among the project beneficiaries
Skin and eye diseases	Increased amounts of readily available water will improve hygienic status for the population at large	High prevalence especially among the groups with the poorest living conditions	Curative: limited. Preventative: limited, but health education and small scale community based water supply projects under implementation	Reduced risk, and an opportunity for the population at large
Enteric diseases (e.g. diarrhoea)	Increased amount of readily available water could improve personal hygiene. Contaminated canal water used as drinking water will have a negative impact.	High prevalence, especially for under fives and groups with poor living conditions	Curative: fair Preventative: health education and small scale community-based sanitation and water supply projects under implementation. Problems with maintenance of wells and pumps.	Moderate risk

Table 8.3. Summary of recommendation safeguards and mitigation measures

Measure	Impact on
Free draining hydraulic structures	Schistosomiasis
Deweeding and desilting of irrigation structures	Malaria and schistosomiasis
Night storage pond bypass structure and spillway	Schistosomiasis
Training of farmers in water management	Malaria and schistosomiasis
Pipelining of conveyance canal	Wwd and schistosomiasis
Foot bridges	Schistosomiasis
Reducing eutrophication of the Goora Dam	Schistosomiasis
Promotion of water supply and sanitation facilities	Wwd and schistosomiasis
Survey to improve design of health protective interventions	General health status
Storeroom for agrochemicals	Agrochemical poisoning
Protective clothing for canal maintenance workers	Schistosomiasis
Support to community based health education during project implementation	General health status
Improved drug supply to local health centre during a 10-year period	Malaria and schistosomiasis
Levelling of borrow pits following construction of dam	Malaria and schistosomiasis
Measures to reduce number of STDs during construction of Goora Dam	STDs
One additional environmental health technician	General health status
One microscopist and one microscope	Malaria and schistosomiasis
Improved intersectoral cooperation during project implementation	General health status

Wwd – water-washed diseases

environmental risk factors and the discussion of 'community vulnerability' will include a clear focus on individual and community-related risk factors.

8.10 ACKNOWLEDGEMENTS

We greatly appreciate the supervisory inputs provided by Martin Birley during the HIA of the Mupfure irrigation project and his input to the drafting of the full HIA report. We are also thankful to the Mupfure community and government officers based in Mupfure for providing most of the information we required for the assessment.

8.11 REFERENCES

Chandiwana, S.K., Taylor, P. Chimbari, M. Ndhlovu, P. Makura, O. Bradley, M. and Gondo, P. (1998) Control of schistosomiasis transmission in a newly established small holder irrigation scheme. *Transactions of Royal of Tropical Medicine and Hygiene* **82**, 874–880.

Chimbari, M. Ndlela, B. and Nyati, B. (1992) Bilharzia in a small irrigation community: an assessment of water and toilet usage. *Central African Journal of Medicine* **38**, 12.

Konradsen, F., Chimbari, M., Furu, P., Birley, M.H. and Christensen, N.O. (1997) The use of health impact assessments in water resource development: a case study from Zimbabwe. *Impact Assessment* **15**, 55–72.

Lanoix, J.N. (1958) Relation between irrigation engineering and bilharziasis. *Bulletin of the World Health Organization* **18**(5–6), 1011–1035.

Le Sueur, D. and Sharp, L.B. (1988) The breeding requirements of three members of the *Anopheles gambiae* Giles Complex (Diptera: Culicidae) in the endemic malaria area of Natal, South Africa. *Bulletin of Entomological Research* **78**, 549–560.

Molden, D. (2007) Water for food, water for life: a comprehensive assessment of water management in agriculture. Earthscan, London.

9

Nam Theun 2 hydroelectric project, Lao PDR

Gary R Krieger, Marci Z Balge, Soutsakhone Chanthaphone, Marcel Tanner, Burton H Singer, Lorna Fewtrell, Surinder Kaul, Pany Sananikhom, Peter Odermatt and Jürg Utzinger

9.1 INTRODUCTION

The Nam Theun 2 hydroelectric project (Figure 9.1) includes the development, construction and operation of a large hydroelectric scheme in Lao PDR in Southeast Asia. Lao PDR, a landlocked country covering an estimated surface area of 236,800 km^2, has some mountainous areas, but the large majority of the country is located within the lower watershed of the Mekong River, which borders or traverses much of the length of the country in a generally north to south direction.

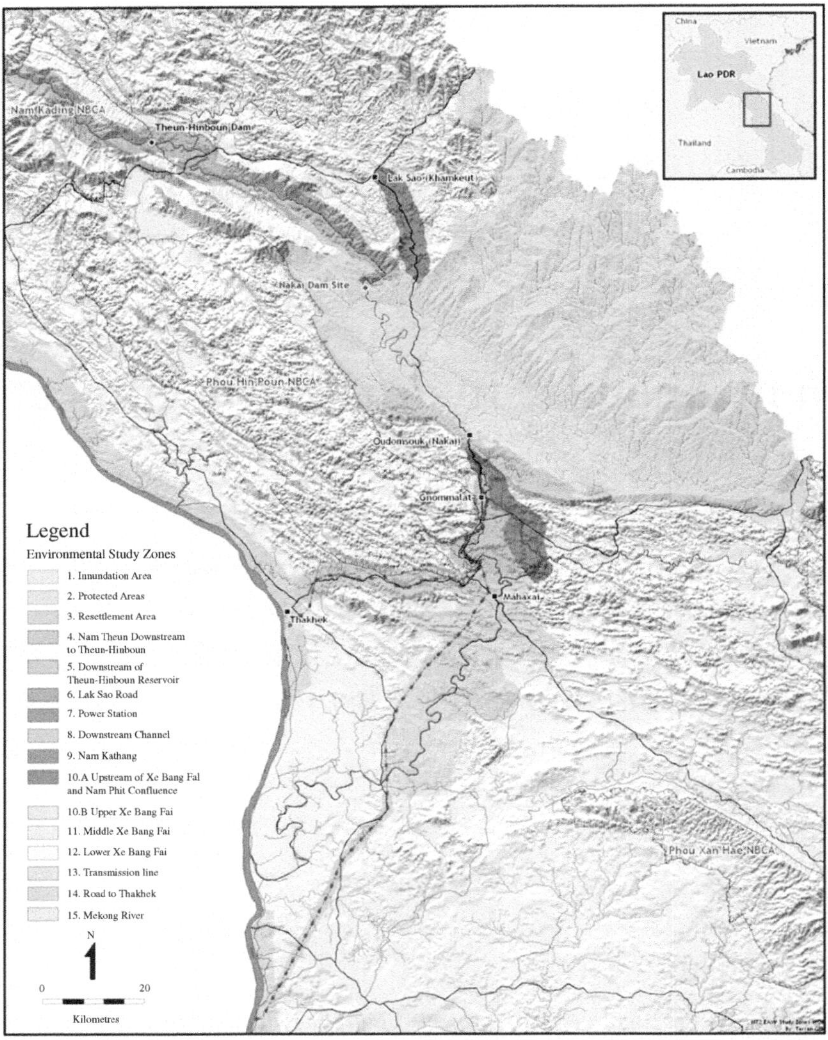

Figure 9.1. Environmental study zones of the Nam Theun 2 hydroelectric project in central Lao PDR (source: Nam Theun 2 Power Company)

In 2006, the estimated population of Lao PDR was 5.8 million (World Bank 2007). Almost 80% of the population currently live in rural areas, in sparsely distributed villages, where access can be problematic due to difficult terrain and the monsoonal climate. The Nam Theun 2 hydroelectric project is located in Khammouane province, central part of Lao PDR, and consists of a main dam on the Nam Theun, a tributary of the Mekong River, creating a 450 km^2 reservoir on the Nakai plateau. Water from the reservoir will drop approximately 350 m to a power station, located at the base of the Nakai plateau. The water discharged from the power station will then flow into a regulating pond and from there to the Xe Bang Fai River via a 27 km-long downstream channel. Hence, the project will divert discharge stored on the Nakai plateau from the Nam Theun to the Xe Bang Fai, producing electricity by utilizing the maximum difference in elevations between the Nakai plateau and the plain.

The government of Lao PDR (GoL) and international development banks consider hydroelectric power exploitation to be of key importance in achieving the country's development goals (GoL 2007). Indeed, the GoL has publicly stated the desire to become the "battery of Southeast Asia". At present, the Nam Theun 2 hydroelectric project is one of the largest infrastructure development projects in Southeast Asia. In 2005, after a preparatory phase lasting over 10 years, the project was approved and financial close was set. Major construction works are underway, with completion of the construction phase due in 2009.

Despite promising economic objectives, large hydroelectric projects have a history of adverse impacts on local populations and ecosystems as a result of profound demographic, ecologic and socioeconomic transformations, including direct, indirect and cumulative health effects (Montgomery *et al.* 1973; Jobin 1999; Lerer and Scudder 1999; McCully 2001; Scudder 2005; Steinmann *et al.* 2006; Erlanger *et al.* 2008a). Whole landscapes are invariably changed with subsequent alterations in the transmission of infectious diseases, since life cycles are closely linked to the environment. Microeconomic shifts, such as changes in household income, directly and indirectly impact health and quality of life at both household and community level. Psychological stress and accompanying mental illness has been an important, yet largely neglected, human consequence of large dams. In many instances, significant numbers of people and their livestock have to be resettled, which can cause social disruption and conflict and result in changes in livelihood (Montgomery *et al.* 1973; Jobin 1999; Scudder 2005; Li *et al.* 2007). Reservoirs can create suitable habitats for snails that act as intermediate hosts for food-borne trematode infections and schistosomiasis (Jobin 1999; Keiser and Utzinger 2005; Steinmann *et al.* 2006) and vectors that transmit malaria (Keiser *et al.* 2005a, 2005b; Lautze *et al.* 2007). The enhanced access and subsequent influx of new residents and temporary labour force accompanied by family members and camp followers to construction sites can alter the frequency and transmission dynamics of sexually-transmitted infections, including HIV/AIDS (Krieger *et al.* 2004). Health impact assessment (HIA) is a powerful approach to predict potential health impacts. Typically, an HIA entails five stages, namely (i) screening, (ii) scoping, (iii) risk assessment, (iv) decision making, and (v) implementation, coupled with monitoring and surveillance (Kemm and Parry 2004). By necessity, HIA must be done prospectively, and it aims to enhance positive health effects and to mitigate negative health effects (Lerer and Scudder 1999; Mercier 2003; Bos 2006; see also Chapter 1).

This chapter reviews the HIA that was performed for the Nam Theun 2 hydroelectric project in early 2004, shortly before financial close of the project. Screening and scoping of the HIA are briefly mentioned. Emphasis is then placed on risk assessment, with key predictions for three selected 'potential impact areas of concern' (PIAOCs) presented. The risk assessment influenced decision making, and hence some of the proposed mitigation strategies are outlines. Additionally, early observations made and experiences gained during the first two years of project implementation (2005–2007), as part of monitoring and evaluation, are briefly mentioned to underscore some of the predictions made during the HIA. Finally, a short post-script is given based on an interim evaluation (carried out in mid-2007) and some of the lessons learned as project implementation moves ahead.

9.2 IDENTIFICATION OF HEALTH ISSUES

The process used to identify and assess the potential impacts on health involved:

- an in-depth review of available regional, national, provincial and district health data;
- analysis and synthesis of project study area-specific baseline economic, health and nutritional surveys;
- comparison of study area data to national, provincial and district health data;
- field survey visit by the core HIA study team; and
- consultation with relevant health stakeholders, particularly Ministry of Health (MoH) officials at the national, provincial, district and local level, and regional representatives from the World Health Organization (WHO) and the World Bank.

The focus of the HIA was on both general and specific public health issues, e.g. water and sanitation (general), malaria and HIV/AIDS (specific). Typical occupational health concerns, such as musculoskeletal injuries, were viewed as 'inside the fence', and hence the realm of the Construction Safety Plan rather than the HIA. Experience with other large infrastructure projects in developing countries (Jobin 2003; Utzinger *et al.* 2004, 2005) demonstrates that the HIA should adopt a reasonably broad perspective when addressing the potential public health impacts and include not only the immediate construction areas but also communities and villages within the transportation and transmission lines sphere of influence and the health impact of individuals who may follow the project (i.e. 'camp followers', who may include both family and relatives of project workers and also service industry providers economically attracted to the project).

9.3 ASSESSMENT PROCESS

Not all of the potential issues in the HIA are clinical medical problems. Potential impacts can also relate to access to health care facilities, socioeconomic factors and cultural practices, e.g. traditional medical providers, indigenous drugs and practices. Hence, a purely biomedical model will not necessarily capture the full range of potential impacts. It is apparent that some important health issues are health determinants that relate to the social and psychological environment (Montgomery *et al.* 1973; Scudder 1973; Cliggett *et al.* 2007). Typically, there has been a tendency to focus on the biophysical environment via a life-cycle model of disease while downplaying key social determinants such as economic status, cultural practices and psychological effects associated with cultural disruption, most notably resettlement (WHO 2000). The HIA of the Nam Theun 2 hydroelectric project made an attempt to incorporate critical psychosocial and culturally important health determinants into the overall assessment of potential impacts.

The project has multiple factors that make the efficient and accurate assessment of potential health impacts a complex undertaking:

- Widespread geography involving multiple types of terrain and habitats.
- Linear feature issues due to construction of large facilities, e.g. reservoirs, roads, transmission lines that cross different terrains, habitats and communities.
- Different types of receptor groups (e.g. construction workers, camp followers, resettlement villagers, etc.) and vulnerable groups within communities (e.g. children under the age of five years, pregnant women and elderly) distributed across time and space.
- Timing issues (e.g. construction mobilization, actual construction phase, reservoir filling, operations and so on) associated with different stages of construction that impact different receptors and locations.

- Seasonality considerations (i.e. rainy versus dry season), where there are potentially significant differences in agricultural/cultivation practices, water use and associated disease transmission dynamics.
- Variability of existing health care infrastructure across different project areas.
- Coordination and alignment with existing national disease control programmes and strategies (e.g. HIV/AIDS, tuberculosis (TB) and malaria).

While these factors can be separately listed and described, it must be understood that they are simultaneously present such that there are direct, indirect and cumulative interactions producing increment, antagonism, synergy and even potentiation of effects or impacts. Given the scope and magnitude of the Nam Theun 2 hydroelectric project, the potential 'universe of outcomes' is virtually limitless. Therefore, in order to impose some type of rational structure on this universe of outcomes, the HIA considered analysing impacts through three different filters.

- Health issues as defined by WHO, namely (i) communicable diseases, (ii) non-communicable diseases, (iii) accidents and injuries, (iv) malnutrition, (v) psychosocial effects, and (vi) social wellbeing (WHO 2000).
- Key sectors as defined by the World Bank, including (i) housing and urban development, (ii) water, food and sanitation, and (iii) transportation, communication and information access and associated linked environmental health areas (EHAs). These include respiratory diseases, vector-borne diseases, sexually-transmitted infections, water-borne diseases, accidents and injuries and exposures to potentially hazardous materials (Listorti and Doumain 2001).
- Disease or chemical specific agents of concern. These include malaria, dengue, soil-transmitted helminthiasis, opisthorchiasis, schistosomiasis, Japanese encephalitis, leptospirosis, acute respiratory infections, TB, sexually-transmitted infections including HIV/AIDS, gastroenteritis (bacterial and viral), psychological disturbances (including substance abuse), malnutrition, trauma/accident-related, air pollution (indoor and outdoor), pesticides, organic and inorganic fertilizers, etc.

Clearly, there is overlap and redundancy across these three filters. In addition, institutional health infrastructure and management capability at the local, district, provincial and national level must be considered. In order to minimize redundancy, while capitalizing on the fact that many of the specific disease agents can be easily classified under broader headings (e.g. respiratory, vector-borne, etc.), the HIA expanded and modified the World Bank environmental health approach by both increasing and broadening the linkage between sectors and broadly defined environmental health. The result of this effort was the creation of eleven EHAs, namely:

(1) Respiratory diseases, including acute respiratory infections (bacterial and viral), pneumonias and TB.
(2) Vector-related, including malaria, dengue and Japanese encephalitis.
(3) Sexually-transmitted infections, including HIV/AIDS, genital ulcer disease, syphilis, gonorrhoea, *Chlamydia* and hepatitis B.
(4) Soil- and water-borne, including soil-transmitted helminthiasis, leptospirosis, schistosomiasis, meliodosis and cholera.
(5) Food and nutrition, including stunting, wasting, micronutrient deficiencies, changes in agricultural practices, gastroenteritis (bacterial and viral) and opisthorchiasis.
(6) Accidents and injuries, including traffic and road-related, construction (home and project-related) and drowning.
(7) Exposure to potentially hazardous materials, including insecticides, pesticides, inorganic and organic fertilizers, road dusts, air pollution (indoor and outdoor, related to cooking, traffic, or other forms of combustion/incineration), landfill refuse or incineration ash, paints, oils or cleaning agents, etc.
(8) Psychosocial (social determinants of health), including violence, security concerns, substance abuse (i.e. alcohol, tobacco and other drugs), depression and communal social cohesion and traditional medicine providers and practices.
(9) Cultural health practices, including traditional medicine providers, indigenous drugs and behavioural practices (e.g. post-partum 'lying by the fire', local term '*yuu fai*').
(10) Health services infrastructure and capacity, including physical infrastructure, staffing levels and technical capabilities of health care facilities at local, district and provincial levels.
(11) Programme management delivery systems, including coordination, harmonization and alignment of the project to existing provincial and national level health programmes (e.g. control programmes of malaria, TB and HIV/AIDS and revolving drug fund) and future development plans. For example, at the national level, there is both an extensive health development plan covering the years 2000 Operation 2020 health goals, namely (i) "Health strategy up to the year 2020", and

(ii) a large World Bank health infrastructure and development project "Health services improvement project" currently at the late development/early implementation stage.

Two additional EHAs, namely (i) non-communicable diseases (e.g. hypertension, diabetes, etc.) and (ii) veterinary health/zoonotic diseases, have been added recently and were considered during the interim follow-up evaluation carried out in mid-2007.

9.3.1 Potential impact areas of concern (PIAOCs)

The project was divided into discrete PIAOCs for HIA purposes. In this concept a PIAOC can represent either a 'subject matter', usually composed of a specific population group (e.g. camp followers) or a defined geographical area (e.g. resettlement area) where project-related health impacts have a reasonable likelihood of occurrence. The 10 PIAOCs are summarised below:

(1) Villages in the Nam Theun 2 watershed: An estimated 5500 people live in some 30 scattered villages located in the forests of the Nam Theun watershed area, covering approximately 3500 km^2.

(2) Plateau resettlement area: Approximately 5880 individuals in 1174 families reside in the zone that will be inundated; hence, the affected population will require relocation to a newly built-up area. A stretch of land along the rim of the Nakai plateau has been designated as the setting where residents of the inundated areas will be relocated (estimated surface area is 208 km^2). The region is vegetated with secondary forest, shrubs and grasses. Some agricultural activities are pursued and will be intensified. A pilot resettlement programme, involving 37 families, commenced in late 2003. This pilot programme provided an opportunity to investigate the direct and indirect impacts of resettlement on changes and adaptations in livelihood activities, primarily intensified farming activities, and health, which should facilitate to readily adapt future mitigation strategies as relocation moves along.

(3) Nam Theun riparian and watershed area: This area includes a small stretch of riparian land from the Nam Theun between the Nakai dam and the Theun Hinboun dam, covering approximately 130 km^2. The watershed has an estimated surface area of 4060 km^2.

(4) Construction work camps: There are four general zones that were identified for the development of construction work camps within the project area. These zones, and their expected peak workforces are:

(i) dam area construction camp (800 workers), (ii) Nakai work camp (600 workers), (iii) power station work camp (2200 workers), and (iv) downstream work camp in the Xe Bang Fai area (400 workers).

(5) Camp followers – family: There is a significant potential that substantial numbers of the workforces' family will co-locate in the same geographical area as the working spouse.

(6) Camp followers – service: It is reasonable to assume that significant numbers of food, entertainment and service vendors will try to co-locate in close proximity to the work camps and/or family followers. Based on recent experiences at other construction sites in central Lao PDR, the overall estimated total camp followers (both family and service) are 2–4 times the workforce census.

(7) Villages along the Nam Kathang and the upper Xe Bang Fai: This area consists of (i) the power station, (ii) the regulating dam site, (iii) the downstream channel area, (iv) Nam Kathang River, and (v) the upper Xe Bang Fai area. An estimated 60 households will be affected by the construction of the downstream channel, and more than 20 villages are located along the Nam Kathang and the upper Xe Bang Fai between the confluence with the Nam Kathang and the Sayphou Xoy Ridge.

(8) Villages along the middle and lower Xe Bang Fai: Twelve villages are located along the middle portion of the Xe Bang Fai. Numerous, fairly densely populated villages are located along the 70 km stretch of the lower Xe Bang Fai that ultimately discharges into the Mekong River.

(9) Transmission line corridor: A 500 kV double-circuit line and a 115 kV single-circuit line will be constructed in order to deliver power produced by the hydroelectric scheme, and primarily exported to neighbouring Thailand. Indeed, the 500 kV power line will be 138 km and run to the Thailand border north of the town of Savannakhet. The 115 kV power line will be 70 km in length and run to Thakhek, the capital of the Khammouane province in Lao PDR. There are villages located within each of these transmission corridors. In addition, there will be moving work camps required for infrastructure installation of the power lines.

(10) Transportation corridor: As part of the project, a variety of new roads and bridges will be constructed and the existing transportation corridor upgraded. Additionally, there will be significant overland truck utilization of existing road networks in order to supply the various construction activities across the entire project.

Each of the 10 PIAOCs was subjected to HIA and respective risk profiles were constructed, showing the 11 EHAs on the *Y*-axis with project timing, at risk population and risk potential on the *X*-axis, using the following legend (Table 9.1).

Table 9.1. Risk profile table legend

	Population affected	Impact level
A	Children/infants < 5 years	High
B	Children aged 5-13 years	Medium
C	Women of reproductive age	Low
D	Men aged 14-60 years	None
E	Elderly > 60 years	Enhanced +

Overall ratings were based on a modified Delphi approach (Rowe and Wright 1999). In early 2004, towards the end of the 2-week project visit of the HIA team, each core member of the team performed an independent rating, followed by discussion and consensus agreement of an overall rating. Subsequently, within a few days, input from key stakeholders and external reviewers was also considered and included. A key feature of these risk profiles therefore is their predictive nature. Whether or not the predictions hold for any given EHA in a specific PIAOC is subject to monitoring and interim evaluations. In this connection it should be noted, however that interventions aimed at mitigating the predicted negative health outcomes are likely to alter health outcomes, and hence our predictions. In any event, this type of exercise – risk profiling by means of a Delphi approach and predictive forecasting – invariably and unavoidably involves subjective professional judgement, and there is always room for debate and disagreement over the relative ranking of impacts. Since predictions of negative health impacts are usually addressed by mitigation strategies, it is difficult to compare the predictive risk profiles with actual occurrences. Therefore, the predicted impact ratings should be considered in broad terms across the project before actual project implementation.

9.4 SELECTED ASSESSMENT RESULTS

In this section the results, along with specific discussion and rationale highlighting some of the most significant potential impacts for the selected PIAOCs, are presented. The nature of these articulations is thus predictive. Indeed, with the data and information at hand as of early 2004, positive and negative health impacts in any of the 11 EHAs were made and stratified for distinct age-groups and PIAOCs. In some instances, our understanding of the predicted risk profiles has been deepened as the project moved along, facilitated by project monitoring and evaluation, including an external interim appraisal carried out in mid-2007. Due to space constraints only three PIAOCs have been selected, namely (i) the Nakai plateau resettlement area, (ii) the construction work camps, and (iii) the villages along the Nam Kathang and upper Xe Bang Fai.

9.4.1 Nakai plateau resettlement area

The most significant potential negative impact relates to the changing agricultural practices/livelihoods, while the most significant enhanced impact relates to multiple sectoral improvements (e.g. improved housing, enhanced access to clean water and improved sanitation) with consequent health benefits. Multiple psychosocial and cultural health impacts have been predicted, which could be significant (adverse) and/or beneficial. Plate 9.1 depicts typical housing conditions on the Nakai plateau before relocation of villages that will be inundated.

Plate 9.1. A typical scene showing part of a village on the Nakai plateau that will be inundated due to the Nam Theun 2 hydroelectric project (Photo: J. Utzinger, 2004)

During the initial construction phases of the project, impacts in all 11 EHAs will be significant (Table 9.2). However, these impacts will be a mixture of positive and negative effects. There should be substantial improvement across all four sub-sectors (housing and urban development; water, food and sanitation; transportation and communication; and information access). The sectoral improvements should directly affect burdens of disease across many of the EHAs (respiratory; vector-borne; soil/water and potentially food nutrition). These positive impacts can only be achieved by significant efforts in environmental areas (health infrastructure), as well as programme management; hence, a detailed implementation plan is essential (as described in the Social Development Plan) (NTPC 2004). In addition, establishment of a rigorous system for monitoring and surveillance of health and social wellbeing is essential.

Table 9.2. Risk profiling of the Nakai plateau resettlement area

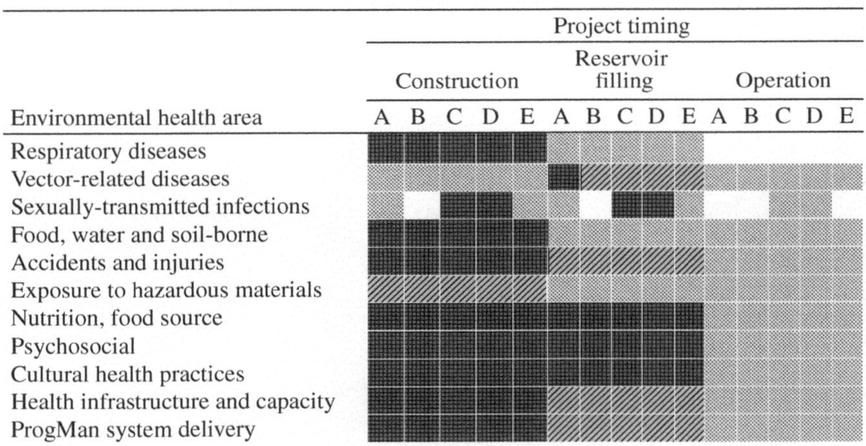

A: Children/infants < 5 years B: Children aged 5–13 years
C: Women of reproductive age D: Men aged 14–60 years
E: Elderly > 60 years ProgMan: Programme management

The potential for negative impacts clearly exists, particularly related to EHAs 3 (sexually-transmitted infections) and 5–8 (accidents; exposures to hazardous materials; nutrition/food). Interaction and mixing with construction workers and/or camp followers could adversely affect sexually-transmitted infections and HIV/AIDS transmission rates. Psychosocial effects of relocation could be significant and more difficult to manage than currently perceived (Montgomery *et al.* 1973; Scudder 1973, 2005; Lumsden 1993; Desjarlais *et al.* 1995). As part of the relocation, there will be required changes in agricultural cultivation practices and livelihoods. These changes may have serious impacts on food production and subsequent nutritional and micronutritional effects. Initially, given the tremendous effort currently underway in this area we expect improvement; however, it is unknown if the proposed change in production/cultivation habits will either work in the intermediate term or be truly sustainable in the long term. A severe challenge to sustainability arises from population growth in the resettled community. Indeed, it is unclear whether the resettlement area which is designed for the current population can handle the resource and food demands of a growing population in the future.

Given the poor soils on the Nakai plateau, increased use of organic and inorganic fertilizers is inevitable. Similarly, the use of pesticides will probably increase. The augmented use of fertilizers, in turn, can have significant short- and long-term impacts on water quality by increasing nitrogen loads via both surface water runoff and discharge to the reservoir and movement into groundwater.

During the full construction phase there will be potentially significant changes (particularly in the dry season) on air pollution levels from vehicle emissions with potential respiratory disease impacts. Two mitigation measures will be implemented to reduce potential respiratory diseases due to road traffic dust, namely (i) permanent watering of unpaved roads during the construction phase, and (ii) locating of resettlement areas at a safe distance away from the main road. Moreover, the new houses provided to the resettlers are larger than the old ones, and occupants living in the houses usually cook under the house or in an improvised kitchen by the side of the house. It is encouraging to note that a review of recent data by the Nam Theun 2 Power Company (NTPC) health team from district hospitals in the project area have not shown any rising trends for consultations related to respiratory diseases.

Road traffic will increase with subsequent increased risk of accident and injury. Indeed, recently collected data by district health personnel reveals an increase in road traffic accidents. However, the majority of accidents are due to villagers using motorbikes rather than the vehicles connected with the construction work. Mitigation strategies are underway; they include road safety education of project staff and enhanced police controls on the roads.

If there is improvement across major sectors (as projected in the Social Development Plan) then overall vector-borne and soil-transmitted and water-borne diseases should significantly decrease. There is a concern that rodent-related diseases (e.g. leptospirosis and typhus) could increase during active construction due to changes in local habitat; however, these potential impacts are felt to be low based on the current experience in the pilot resettlement village. Japanese encephalitis has not been documented and is not expected. Nevertheless, full implementation of environmental management plans in addition to active monitoring and surveillance are required for vector-borne diseases.

During the early operations and reservoir filling stage, the possibility of changes in vector-related diseases is real. In the resettlement area, villages will not be in close proximity to forest cover so that contact with the major malaria vectors will be lessened; however, a change in the vector mix (e.g. *Anopheles nivipes* versus *An. minimus*, *An. dirus* and *An. maculatus*) cannot be discounted since villages will be extremely close to the reservoir. Regardless of project phase, the potential for significant disruption in the current communal political, social, cultural and economic relationships across the Nakai plateau exists. If there are substantial and unmanaged disruptions to community cohesion then significant psychosocial impacts including increases in substance abuse, violence and other psychiatric disorders are possible. Impacts to the traditional medicine practitioners are already ongoing and related to the revolving drug fund. The revolving drug fund appears to have significantly altered the reliance on, and utilization of, traditional practitioners, with the notable exception of traditional birth attendants. These changes are currently unrelated to the project but will probably accelerate with better delivery and access to medical care.

Similar impacts are expected on post-partum medical practices. The current post-partum 'lying by the fire' (*yuu fai*) practice is complex and deeply rooted in the culture. Traditionally, after delivery, Lao women remain for 15–30 days on their bed under which a fire with charcoals is made. A strict dietary regime of rice soups and chicken breast meat is followed. If this dietary practice is rigidly adhered to for long duration, the women are at risk of anaemia and micronutrient deficiencies. The newborn could be similarly impacted.

At the University Hospital Mahosot in Vientiane, 16% of all admissions between 1990 and 2000 were due to infant beriberi. The importance of this practice and the level of vitamin B1 deficiency in the project area are unknown. Despite the negative potential impact, the current practice has many positive attributes, e.g. family cohesion, socialization, supportive role of the male spouse and other direct family relatives. The post-partum duration appears to be highly flexible ranging from 2 to 6 weeks. In addition, the nutritional and caloric composition of the diet also appears to be highly variable. Based on the core HIA team's observations, both the nutritional practices and the length of the lying by the fire were highly variable and potentially dependent on the specific ethnicity of the woman. Additionally, recent data from a Mahosot Hospital study (Soukaloun *et al.* 2003) demonstrated that both rice cooking practices, food selected by the lactating mother and the family's socioeconomic status were highly related to the rate of clinically significant thiamine deficiency in breast fed infants in Lao PDR. Thus, the potential adverse effects of the *yuu fai* may be amenable to educational intervention strategies. The potential impact of the project on this post-partum practice is unknown. The opportunity exists to improve and reinforce the beneficial aspects while ameliorating the negative potential nutritional aspects via educational outreach. Whether relocation and 'urbanization' will change the overall practice is unknown since this practice already flourishes in urban areas, including the capital city of Vientiane. This is one of many instances where the imposition of long-term monitoring of impacted communities could allow for a meaningful assessment of the consequences of the project. It could also provide a basis for adaptive assistance and community self-management that could facilitate improved living standards.

The overall impacts at the district level on EHAs 10 and 11 will be significant. However, these impacts could be quite positive if overall capacity, management and delivery systems are improved via enhanced funding and coordination with other national and provincial level efforts.

9.4.2 Construction camps

The development and presence of multiple construction camps during the various phases of the project will be an obvious source of both enhanced and adverse impacts. While most of the workers will probably be males, the adjacent community interactions and potential impacts are expected to occur across all ages and both genders. The most significant potential negative impacts are (i) transmission and potential amplification of sexually-transmitted infections, including HIV/AIDS; (ii) secondary developments triggered by camp followers living in close proximity to construction work camps; and (iii) overwhelming of local management and programme management systems. On the other hand, the presence of the construction work camps will clearly be a major economic stimulus that in turn acts as a huge health determinant. Consequently, the presence of the camps can act as a significant catalyst for both health care delivery infrastructure and technology enhancement.

Potential significant adverse impacts could occur across all environmental areas (Table 9.3), however, most of these potential impacts will probably be time limited and directly related to the presence or absence of the camps. In contrast, changes in the shape and slope of the HIV/AIDS epidemic curve would be profound and long lasting. The current level of HIV/AIDS in the major project province (Khammouane) is extremely low. Factors that change mixing rates (interaction between high and low risk groups with disparate seropositive prevalence rates) and partner change rates (increased disposable spending

Table 9.3. Risk profiling of construction work camps

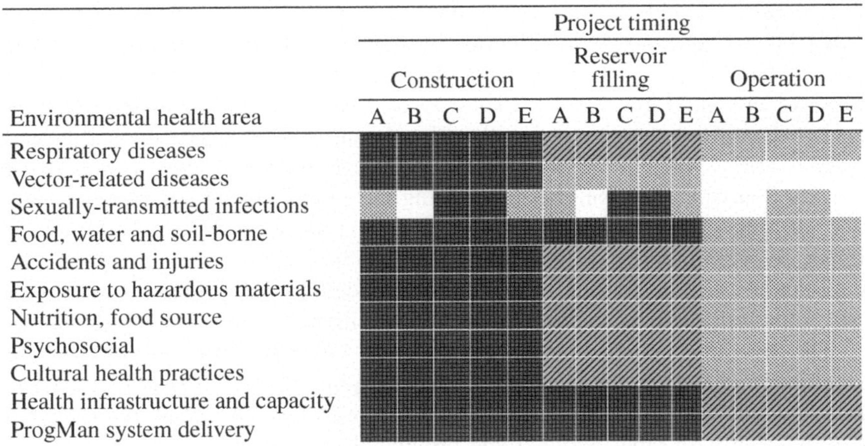

A: Children/infants < 5 years B: Children aged 5–13 years
C: Women of reproductive age D: Men aged 14–60 years
E: Elderly > 60 years ProgMan: Programme management

money leading to increase purchase of commercial sex) directly affect the shape and size of the epidemic curve (Krieger *et al.* 2004). This situation is often characterized by four "M's":

- men;
- money;
- mixing, and
- movement.

The construction work camps clearly have all of these parameters in simultaneous interaction. Therefore, a well developed, monitored and sustained multi-component HIV/AIDS programme is of paramount importance. Given the low baseline levels, the HIV/AIDS impacts are so potentially significant that the development, review, funding and coordination (or direct management) with/by the national programme should be an absolute requirement prior to construction camp development. The development of a formal HIV/AIDS impact assessment is a possibility; however, it is probably a better utilization of resources to fully concentrate on a multi-level programme since the impact assessment will simply document the obvious.

In terms of the other 10 defined EHAs, the construction work camps can be viewed as an exercise in rapid urban planning with all the well-known and expected positive and negative consequences. There are a myriad of enhanced and negative effects that can and are likely to occur. However, our assessment is that there are several key EHAs that could be significantly (adversely) affected and 'spill-over' into adjacent villages/communities.

9.4.2.1 Respiratory diseases

Baseline health data obtained in 2000/2001 (not shown), suggested that TB was undoubtedly under-diagnosed. In addition, there was no formal case finding system in place at the time of the current HIA in this study area and elsewhere in Lao PDR; hence, prevalence rates are likely to be higher than reported (Odermatt *et al.* 2007; Erlanger *et al.* 2008b). TB and HIV/AIDS are covariate with each other in almost all epidemiological settings, i.e. in sub-Saharan Africa approximately 70% of the TB cases are HIV positive. While the sub-Saharan African situation may not be directly transferable to Lao PDR, it is nevertheless a cautionary experience that will require close monitoring and surveillance. The current health infrastructure and management systems capacity for TB diagnosis, treatment and case finding and surveillance would be overwhelmed by a dramatic upsurge in TB regardless of the trigger.

9.4.2.2 Vector-related diseases

Significant progress has been made regarding the control of malaria in the project areas and elsewhere in Lao PDR over the past several years, through a combination of village-level early diagnosis (symptom-based) and treatment (via the revolving drug fund) and the large-scale implementation and utilization of insecticide-treated nets (ITNs) (Trung *et al.* 2004). At the time of the current HIA in early 2004, chloroquine was used as the first-line drug despite repeated reports of high levels of resistance across the country. Since then, chloroquine is only used in the treatment of *Plasmodium vivax* malaria. Meanwhile, the national malaria control protocol has been revised and, at all levels, artemisinin-based combination therapies (ACTs) are used. In fact, artemether-lumefantrine (Co-artem) serves as the first-line antimalarial therapy.

Construction activities may alter ecosystems in unanticipated ways and change the pattern, distribution and dominant vector species. A comprehensive malaria control programme is essential. Implementation of such a programme as part of the provision of systemic health services, including prevention measures, would require a commitment by the GoL that runs well beyond the time frame of designated project completion. A collaborative approach by NTPC and Institut

Pasteur has a long-term monitoring programme of project impacts on the resettled populations. A number of initiatives are underway and include diagnosis of malaria, dengue and Japanese encephalitis and any emerging infectious diseases. The collaboration also monitors malaria vector abundance, spatially and temporally, in order to enhance integrated vector management.

9.4.2.3 Food, water and soil-borne diseases

There is significant potential for rodent control problems associated with both construction camp development and 'steady-state' operation. There are detailed environmental management plans for this activity, however, if these plans are not fully and successfully implemented then rodent-related diseases in adjacent villages/communities could become significant (e.g., leptospirosis, murine and scrub typhus).

While the project construction camps may provide a significant positive market-based signal for enhanced agricultural production, price inflation both at a macro level (e.g. unit of the camp) and at a micro level (e.g. individual construction worker purchasing goods from nearby markets) is possible. The least productive members of the community could be nutritionally marginalized if there is significant food price inflation.

Opisthorchis viverrini infection prevalence rates currently are low in the Nakai plateau area, but extremely high in the lower regions of the Xe Bang Fai. Movement of workers infected with *O. viverrini* into the Nakai plateau area with subsequent introduction of the parasite into local waters and fisheries and introduction of different dietary practices could change transmission rates considerably.

9.4.2.4 Accidents and injuries

There will be significant numbers of both light and heavy vehicles entering and leaving the construction areas (Plate 9.2). The likelihood of accident and injury with local residents will significantly rise simply as a function of increased traffic and activity. Consequently, we expect, regardless of the implementation of a required traffic management plan, a significant (adverse) rise in traffic-related injuries, accidents and fatalities with subsequent pressure on health care infrastructure and delivery systems.

9.4.2.5 Exposure to hazardous materials

While there is some potential for community/village level exposure to project-related hazardous materials, this is considered low. In contrast, potentially significant exposure to increased levels of air pollution is a certainty. Overall air

Plate 9.2. Road transportation will significantly increase in the Nam Theun 2 hydroelectric project area, and thus increase the risk of accidents and injuries (Photo: J. Utzinger, 2004)

pollution will increase as a direct function of 'vehicle kilometres travelled' (VKT). The planned level of construction will significantly increase VKT in the various construction areas. Since many of the roads will not be paved, an increase in fugitive dusts is a certainty (Plate 9.2).

Similarly there will be construction-related (earth works) fugitive dust releases. Road-related fugitive dust releases typically have varied particle sizes. Particle size distribution curves from mechanical forces (e.g. tyres) versus combustion (e.g. engines, incinerators, etc.) are quite different. Mechanical fugitive dusts tend to have larger and less respirable particles; however, there will be an overall increase in particulate matter (PM) of 10 microns (PM 10) or less, including smaller and toxicologically more significant particles less than 2.5 microns (PM 2.5). Since the impact of dust is more pronounced on the construction workforce than the general population, mitigation measures need to be implemented by the workforce contractors rather than the district public health teams.

9.4.2.6 Psychosocial (social determinants of health)

The Social Development Plan considers both the numerous psychosocial impacts that could be triggered by the project and the available mitigation strategies for these impacts. The introduction of the construction camps and large numbers of workers with disposable money is potentially associated with community/village level changes in existing substance abuse rates and adverse interaction with local members of nearby communities. Strict alcohol and drug control policies and procedures are essential. Smoking prevalence rates, using home grown or inexpensive national or foreign cigarettes, appear to already be quite high so it seems less likely that these rates will be significantly impacted. Labour trafficking issues are unlikely.

9.4.2.7 Cultural health practices

Medical care will be provided to workers as part of the overall head contractors health care programme. Based on the experience with the revolving drug fund, modern health care quickly drives out traditional practices, particularly if the cost of the health care is very low. Males will probably be the majority of the workforce; hence, post-partum issues are not likely to be immediately significant. However, the disruption caused by remote site work is unknown. For local labourers, a full time construction job will clearly impact the '*yuu fai*' role of the male spouse (see 9.4.1).

9.4.2.8 Health infrastructure and capacity

Even if the construction camps were perfectly self-contained, there will clearly be increased demands placed on the existing health care delivery infrastructure. As noted before, there will be an inevitable increase in road-related accidents and injuries, which will involve community members. Regardless of the overall level of control, some upsurge in camp followers is to be expected. Trauma-related injuries place complex and significant demands on the medical care infrastructure in equipment, training, resupply and system management.

9.4.2.9 Project management system delivery

The potential demands on existing provincial and national level major programmes will be significant. Project-related efforts must be fully aligned and coordinated with current efforts. There is an opportunity for enhancement of existing efforts but it will require detailed planning, communication and coordination.

9.4.3 Villages along the Nam Kathang and upper Xe Bang Fai

The major impacts in this PIAOC are related to (i) changes in agricultural practices and fishing yields due to increased year-round availability of relatively large volumes of water, and (ii) construction activities in areas adjacent to population centres (Table 9.4).

Table 9.4. Risk profiling of villages along the Nam Kathang and upper Xe Bang Fai

	Project timing														
	Construction					Reservoir filling					Operation				
Environmental health area	A	B	C	D	E	A	B	C	D	E	A	B	C	D	E
Respiratory diseases															
Vector-related diseases															
Sexually-transmitted infections															
Food, water and soil-borne															
Accidents and injuries															
Exposure to hazardous materials															
Nutrition, food source															
Psychosocial															
Cultural health practices															
Health infrastructure and capacity											+	+	+	+	+
ProgMan system delivery															

A: Children/infants < 5 years
B: Children aged 5–13 years
C: Women of reproductive age
D: Men aged 14–60 years
E: Elderly > 60 years
ProgMan: Programme management

Enhanced water availability will probably allow for dry season rice cultivation, but may adversely affect fishing yields. Hence, the possible significant impacts – positive or negative – are largely related to the changing water flow dynamics in the Xe Bang Fai River, significantly higher water discharges all year long and altered water quality. Adverse nutritional impacts are primarily related to potential decreases in fishing yields and, probably to a lesser extent, to somewhat smaller areas for river bank gardening (Plate 9.3). These potential adverse impacts are uncertain but assessed as 'significant' since fishing provides an important source of dietary protein for local populations. High water volumes during the dry season could impact river bank gardening practices since there is a limited area for this type of cultivation practice. The highest potential impact rating was not given since it is unknown if these impacts will be experienced and, if so, to what degree, and for what duration. For example, local fishermen are likely to adapt their fishing gear as a result of permanently elevated water discharges.

The vector-related diseases have a moderate likelihood of gaining in importance due to increased water availability. Malaria is already prevalent in this area, however, a change in dominant vector species to *An. nivipes* is considered possible based on the experience seen in Thailand and India. Also, there is some suggestive evidence related to this hypothesis based on the published work of Kobayashi *et al.* (2000). Increased rice field exposure may

Plate 9.3. River gardens along the Xe Bang Fai during low water discharges in the dry season (Photo: J. Utzinger, 2004)

increase meliodosis rates, a bacterial disease associated with rice farming. A second water-resource driven rice cultivation period will increase the need and demand for fertilizers and pesticides. The current Pest Management Plan describes the pattern and utilization of these chemicals in the Xe Bang Fai region. With increased utilization, exposure, residue accumulation and surface water runoff are considered likely. This sequence of events already occurs; it is the degree to which it will be augmented that is unknown.

Opisthorchiasis is highly endemic in villages along the Nam Kathang and upper Xe Bang Fai, but the increased availability of water resources is unlikely to change this situation. Schistosomiasis is not present in the region, however, the intermediate host snail has been documented in multiple locations. Transmission of schistosomiasis mekongi is currently restricted to two districts in the very south of Lao PDR with an estimated 60,000 people at risk (Urbani *et al.* 2002). While there is some possibility of (i) workers emigrating from these endemic areas in the south, (ii) relocating to the project area, and (iii) completing the transmission cycle, the chances of this scenario appear to be low. In addition, it has been suggested that the increased overall water flow levels create a less favourable habitat for the intermediate host snails.

While there may be limited relocation of some households along the downstream channel, the number of affected individuals is small and the relocation distance is also quite minimal. Additionally, unlike for the Nakai resettlement area, a potential change in lifestyle is less likely. From an enhancement impact, the food/nutrition situation could improve by virtue of a second rice growing season, however, there could be a short-term impact during construction and filling phases since there will be a loss of land under cultivation. The psychosocial impacts due to loss of land, change in cultivation practices and accident-related injury and death were rated moderate since these changes will be timing restricted with eventual improvement. Changes to cultural health practices are felt to be moderate. During the construction phase, there will be an increased demand on available health infrastructure, which is facilitated by the improvements in health infrastructure and capacity supported by NTPC. Recent data obtained from ongoing monitoring and surveillance activities suggest that utilization rates have already increased since the start of the project in May 2005.

9.5 MITIGATION

Mitigation strategies for all of the PIAOCs have been proposed, and some pointers have already been given before. Here, mitigation strategies will be illustrated for the Nakai resettlement area only as an example (Table 9.5).

Table 9.5. Post-mitigation risk profiling of plateau resettlement area

	Project timing														
	Construction					Reservoir filling					Operation				
Environmental health area	A	B	C	D	E	A	B	C	D	E	A	B	C	D	E
Respiratory diseases											+	+	+	+	+
Vector-related diseases															
Sexually-transmitted infections															
Food, water and soil-borne											+	+	+	+	+
Accidents and injuries															
Exposure to hazardous materials															
Nutrition, food source															
Psychosocial															
Cultural health practices															
Health infrastructure and capacity											+	+	+	+	+
ProgMan system delivery															

A: Children/infants < 5 years B: Children aged 5–13 years
C: Women of reproductive age D: Men aged 14–60 years
E: Elderly > 60 years ProgMan: Programme management

9.5.1 TB control

TB prevention programmes will be enhanced within the project area through the MoH. All patients visiting the Nakai and Gnommalat district hospitals who are diagnosed with TB will receive case management follow-up by a district health worker. These follow-up activities will include visits to the patient's residence to interview family members and close relatives and conduct interviews and skin testing as appropriate to determine the likelihood and extent of TB transmission. In addition, it is proposed that district health worker will implement the directly observed therapy, short-course (DOTS) strategy with follow-up through the village health volunteer in charge of the revolving drug fund. Additionally, better housing and nutrition in the resettlement area are likely to reduce the incidence of TB as experienced in Western Europe after the industrial revolution.

9.5.2 Housing design

Housing design is specifically linked to respiratory diseases. Smoke from cooking and heating with wood, charcoal and kerosene irritates mucous membranes of the respiratory tract. Poor ventilation inhibits the dispersion of droplet spread infections and lack of sunshine prevents natural sterilization of the air by sunlight. Health education and awareness campaigns to reduce indoor and outdoor air pollution will be coordinated through the MoH and World Bank. Respiratory infections are transmitted by airborne particles, droplets or physical contact therefore personal hygiene education will be stressed. Community education campaigns to reduce road dust will be conducted. During community education programmes, opportunities will be explored regarding the use of electric burners for cooking rather than indoor or outdoor fires.

Housing, landscape design and land use planning appear to offer diverse sustainable opportunities for reducing human exposure to malaria vectors. Any modification of housing design that affects the density of *Anopheles* mosquitoes that breed nearby or their ability to enter or to leave housing structures may reduce the risk of malaria infection. Malaria risk generally increases with proximity to productive *Anopheles* breeding sites (Subramanian *et al.* 1991; Staedke *et al.* 2003; Oesterholt *et al.* 2006). In some cases, a progressive gradient of risk can be demonstrated in homes adjoining well-defined breeding sources (Trape *et al.* 1992; Thompson *et al.* 1997).

The quality of housing and method of construction have repeatedly been associated with risk of malaria transmission (Lindsay *et al.* 2003, Keiser *et al.* 2005a, 2005b; Kirby *et al.* 2008). In the small town of Kataragama in southern Sri Lanka, for example, the construction quality of homes ranges from 'poor houses' with open-eaves, palm frond roofing and mud walls and floors to 'good houses' with closed eaves, tile roofs, brick exteriors and finished plaster interiors (Gamage-Mendis *et al.* 1991; Gunawardena *et al.* 1998). Malaria risk for those living in the poor houses was three times greater than for residents of the good houses. When exposure was amplified by close proximity to a breeding site, malaria risk became four times greater for those living in poor houses. Living less than 100 m from a breeding site tended to double risk for those living in poor houses. Overall, a tenfold variation in risk existed across all households involved in this study, possibly reflecting the strong effect of domestic microenvironments.

In India, dwellings with open eaves similarly raised the relative risk of clinical malaria for their inhabitants (Subramanian *et al.* 1991). Houses can readily be designed with closed eaves, especially when thatch is not used for roofs. Certain homes with open eaves might be retrofitted with false ceilings to provide a similar exclusionary effect. Inexpensive materials, such as hessian cloth, can sometimes be used towards this end. False ceilings further insulate living spaces from thermal radiation.

The biting habits of the various local vectors of malaria and the proximity of domestic animals should be taken into consideration when designing sleeping arrangements. In New Guinea, for example, it was found that raised sleeping platforms markedly reduce the frequency of mosquito bites in households, because the local vector, *An. farauti* is adapted to quest at ankle height (Charlwood *et al.* 1984). It was suggested that keeping pigs under platforms would further reduce biting rates due to the zoophilic tendencies of the vector. Mixing livestock with human habitation has also been suggested in Ethiopia where *An. arabiensis* is the dominant vector. People residing in mixed dwellings in such areas (Hadis *et al.* 1997) experience contact with only a fourth as many vector mosquitoes as do those without cattle or whose cattle are isolated in cattle sheds. Such a zooprophylactic effect can reduce the force of malaria transmission, with an effect proportional to the square of the resulting decrease in anthropophily. However, to make a significant dent on malaria transmission, mathematical modelling suggests that a large number of domestic animals that are easily accessible for mosquito blood meals are required (Sota and Mogi 1989). Moreover, non-human hosts would not divert strongly anthropophilic mosquitoes. In any event, zooprophylactic measures (assuming that the local mosquito vectors exhibit favourable behaviour) may be quite practical in Lao PDR since this pattern of human livestock cohabitation is already quite common. However, zoonotic diseases might gain in prominence where human and domestic animal contacts were enhanced.

Housing structures within the pilot project – after consultation with local residents – have been designed above the ground, which is a deterrent for mosquito entry into the home. However, eaves have been left open, windows are

Plate 9.4. New houses built in the relocated villages on the Nakai plateau. Of note, houses are raised, there are no window screens, eaves are open, rainwater is collected in closed tanks (for washing and cleaning purposes) and a toilet has been constructed for each household; hence water supply and sanitation are significantly improved (Photo: S. Kaul, 2007)

not screened, and flooring is not tight, permitting multiple mosquito entry points into the structures (Plate 9.4). House screening has contributed to the elimination of malaria from many parts of the world. Additionally, studies have shown that houses with open eaves, or which lack ceilings are associated with increased numbers of mosquitoes and higher levels of malaria compared with houses with closed eaves or ceilings (Lindsay *et al.* 2003). The core HIA team thus recommended that for future construction of resettlement homes, one of two options might be considered:

- Screen the windows and eaves with 16 mesh screening material and construct flooring such that boarding does not have cracks that permit mosquito entry. This is the preferred method.
- Provide insect screen ceilings for each structure, in addition to ITNs. Studies have shown that screen ceilings do not require treatment in order

> to be effective. Therefore, this can be a simple and cost-effective strategy (Lindsay *et al.* 2003). The study cited applies to *An. gambiae*, one of the most efficient malaria vector species of sub-Saharan Africa. It would be interesting to carry out detailed entomological studies in the project area to verify that *An. dirus*, *An. minimus*, *An. maculatus* and other potential malaria vector species in Lao have similar flight patterns, to confirm whether this approach would also be effective in Lao PDR.

It should be noted, however that the effectiveness of these measures will depend on the users of the mosquito-proof houses. Key informant discussions revealed that villagers are currently not used to closing the doors to prevent insects entering the houses. Moreover, a common practice is that villagers sit in the veranda of their homes in the evenings until bedtime, which is the peak time for mosquito bites. Hence, the proposed strategies will require extensive education programmes, and changes in the practice and behaviour of the local communities will take a considerable amount of time.

9.5.3 Vector-borne diseases

Malaria prevention strategies of ITNs and the provision of access to appropriate antimalarial drugs through the village-based revolving drug fund programmes resulted in significant improvements (Kobayashi *et al.* 2004), and hence they will continue within the project area.

Increased focus on public health education will be coordinated through the MoH and World Bank programmes regarding the prevention of dengue through bite prevention and, where appropriate, elimination of mosquito breeding sites of the vector species *Aedes aegypti*. These mosquitoes live and breed in clean, fresh and salt standing water, e.g. water pots, cisterns, small containers, temporary pools and periodic flooding.

If resources allow, during pre-employment exams, applicants should be screened for *Schistosoma mekongi*, *O. viverrini* and soil-transmitted helminth infections and, if need be, treated with praziquantel (current drug of choice for schistosomiasis and the majority of food-borne trematode infections; Keiser and Utzinger 2007) and albendazole or mebendazole (current drugs of choice for soil-transmitted helminthiasis; WHO 2002). This strategy should help avoid the emergence of schistosomiasis via workers from southern Lao PDR to the project area where it currently does not exist. Moreover, helminth infections are negatively associated with worker productivity, and hence deworming is likely to enhance overall work performance (Gilgen *et al.* 2001).

9.5.4 Sexually-transmitted infections

Evidence from other construction projects suggests that work camp followers, including service workers, will set up facilities adjacent to construction camps. Although it is not possible to completely isolate the camps, particularly when they are in close proximity to settlements, it is possible to restrict access to the camps for non-residents. The GoL will monitor and discourage new settlements of camp followers within the project area. Prevention programmes are currently in place and additional mitigation will focus on supporting local measures aimed at reducing incidence.

9.5.5 Food and water-related diseases

Increases in gastrointestinal diseases (e.g. diarrhoea and dysentery) are among the most important potential health impacts. Baseline health data obtained from over 800 households on the Nakai plateau area and more than 1600 households in the Xe Bang Fai downstream area in 2000/2001 showed that the majority of households use unsafe drinking water and only a small fraction of households have access to any kind of improved sanitation (Erlanger *et al.* 2008b). Domestic water supply, latrines and electricity will be provided to all households within the resettlement area (Plate 9.4). Community health education programmes regarding the maintenance of safe drinking water and the management of latrines will be conducted. Basic sanitation measures to prevent faecal-oral transmission of diseases will be stressed. These measures, along with a large-scale deworming campaign, have already resulted in lowering of helminth infections among the resettlers as reported by NTPC health programme.

9.6 CONCLUSIONS

The HIA summarised here was a systematic exploration of the range of potential health impacts, positive and negative, that could be induced by the Nam Theun 2 hydroelectric project in central Lao PDR. There are two inherent limitations of this type of exercise that must be understood: (i) subjective, but critical, professional judgment must be applied, and (ii) for a complex project like this, there is substantial uncertainty regardless of the depth or detail of baseline planning and investigation. With these limitations in mind, there are a number of overall assessments that can be reasonably articulated and scientifically defended:

- both positive and negative health impacts will occur;
- influx, triggered by resettlement and development of large work camps is a key characteristic;

- unintended or unexpected positive and negative impacts are inevitable;
- mitigation strategies, if successfully designed and implemented, can substantially reduce the most likely predicted negative impacts; and
- an objective and rigorous system of long-term monitoring and surveillance is essential and, in our view, the only realistic method of actually determining the 'true' impacts, including direct, indirect and cumulative level effects in the broadest possible manner, including equity, health and social wellbeing.

The most likely series of obvious impacts is inextricably related to the construction workforce and possible associated camp followers (family and service personnel) and the resettlement of over 6000 individuals on the Nakai plateau. In the first instance, when a large force of construction workers, logistical support personnel and mechanized equipment interfaces with a largely underdeveloped population in an intrinsically high disease burden environment, multi-level impacts will occur. These impacts are substantially accentuated if the project attracts large numbers of camp followers to the same limited geographical area. The interaction and mixing of these groups with each other and with the local population in an environment that has a substantial level of disease is the source of the most likely impacts during the initial stages of project implementation and operation. The resettlement area is an obvious source of multi-level impacts, positive and negative, across all EHAs. If the Social Development Plan is successfully implemented, the overall net effect could be positive for the resettlement area group. However, continued population growth must be taken into consideration, as this will significantly impact on the scarcity of natural resources, and hence govern vulnerability and resilience of affected populations. Measures that (i) modify and control the interaction and interface among groups, and (ii) improve the underlying disease burden via improvements in basic sectors (housing, water/sanitation, transportation and communications) will usually tend to mitigate, but not necessarily completely eliminate, impacts.

Other, more subtle, areas of potential project impacts are related to the Xe Bang Fai downstream areas. The risk profiles for the Xe Bang Fai impacts are generally quite different to those associated with the active construction or the resettlement area and are more likely to be experienced (if at all) once the reservoir filling has been completed and the operations phases of the overall project are launched, particularly in the first years of operation until a new steady-state system is found. For the downstream areas, impacts could be quite positive due to enhanced opportunities for irrigation-based food cultivation and production. Conversely, riverbank gardens will be destroyed due to constantly higher water levels and if there are impacts on fishing yields then there could be significant negative consequences on dietary protein intake.

Impacts on psychosocial wellbeing and cultural health practices are inevitable and largely unpredictable in terms of magnitude, potential severity or beneficial effect. There could be substantial positive psychosocial benefits associated with sectoral improvements in the resettlement areas. Conversely, the negative psychosocial effects of major projects in general, and relocation in particular have often been underestimated. Therefore, rigorous monitoring and surveillance for these effects is essential. The modernization of health practices in Lao PDR via the revolving drug fund has already directly and visibly impacted traditional medical practices. The project is likely to accelerate this trend; however, in some circumstances this will be beneficial, e.g. the tremendous improvement in the fight against malaria.

9.7 POSTSCRIPT

At the mid-construction point of the Nam Theun 2 hydroelectric project (evaluation carried out in summer 2007), the overall community public health efforts have achieved and established a high level of activity. In terms of performance, the achievements have been substantial. However, strengthening of the routine data collection and some additional indicators implemented by the NTPC health team will demonstrate over time the level of improvements or deteriorations among the affected populations. A variety of structural and management issues are emerging that could adversely compromise the long-term sustainability and success of the overall health effort. Ownership and adaptability of the government health workers to the public health management plan will be crucial in sustainability of the project during the remaining construction phase and subsequent operation phase. Critical data and diagnostic coordination and communication between the internal health and safety project staff, the provincial and national health authorities, and the internal project-related community health outreach group will require further strengthening.

There is scope for improvement in key programmes such as sexually-transmitted infections and outbreak/epidemic monitoring, which will require further strengthening. As expected, initial health data reviews indicate both positive and negative impacts. It is most encouraging to note that the prevalence and intensity of parasitic infections and indeed multi-parasitism are rapidly dropping across the resettlement villages. On the other hand, influx of construction workers and presence of large work camps has attracted a considerable number of young and potentially vulnerable women working in restaurants and bars. An additional longer term concern is the potential rapid rise in resettlement population fertility rates; therefore, family planning issues and general reproductive health concerns will become increasingly important. Whether or not the land occupied by the resettled community today can support a much larger population in the future is an issue of concern. The maternal and child health activities, including family planning, implemented within the government health infrastructure by NTPC health team should address fertility to some extent. Educational facilities provided are expected to equip the younger generations to seek employment/migration outside the Nakai plateau.

Finally, the overall monitoring and surveillance need to be understood and adhered to by the district health services with further integration into the social monitoring system. Overall, there are many short-term successes that have been achieved and should be applauded. In order to build a structure that will sustain beyond the construction and operation phase, surveillance of key EHAs are warranted so that negative health effects can be further mitigated and positive effects further enhanced.

9.8 ACKNOWLEDGEMENTS

The core study team is grateful to NTPC for having had the chance to carry out the HIA of the Nam Theun 2 hydroelectric project and to summarize this work in the current book chapter. We are especially grateful to Mr. Chris Flint, former chief of the Social Development Plan. Thanks are also addressed to Dr. Robert Bos from WHO for critical review of the initial HIA report and for constant encouragement to write up this work. M. Tanner acknowledges financial support from the National Centre of Competence in Research (NCCR) North-South programme "Research Partnerships for Mitigating Syndromes of Global Change," individual project no. 4 (IP4) "Health and Well-Being. P. Odermatt (project no. NF3270B0-110020) and J. Utzinger (project nos. PPOOB-102883 and PPOOB-119129) are grateful to the Swiss National Science Foundation.

9.9 REFERENCES

Bos, R. (2006) Health impact assessment and health promotion. *Bulletin of the World Health Organization* **84**, 914–915.

Charlwood, J., Paru, R. and Dagoro, H. (1984) Raised platforms reduce mosquito bites. *Transactions of the Royal Society of Tropical Medicine and Hygiene* **78**, 141–142.

Cliggett, L., Colson, E., Hay, R., Scudder, T. and Unruh, J. (2007) Chronic uncertainty and momentary opportunity: a half century of adaptation among Zambia's Gwembe Tonga. *Human Ecology* **35**, 19–31.

Desjarlais, R., Eisenberg, L., Good, B. and Leinman, A. (1995) Dislocation. In *World Mental Health: Problems and Priorities in Low-Income Countries.* Oxford University Press, New York.

Erlanger, T.E., Krieger, G.R., Singer, B.H. and Utzinger, J (2008a) The 6/94 gap in health impact assessment. *Environmental Impact Assessment Review* **28**, 349–358.

Erlanger, T.E., Sayasone, S., Krieger, G.R., Kaul, S., Sananikhom, P., Tanner, M., Odermatt, P. and Utzinger, J. (2008b) Baseline health situation of communities affected by the Nam Theun 2 hydroelectric project in central Lao PDR and indicators for monitoring. *International Journal of Environmental Health Research* **18** (in press).

Gamage-Mendis, A., Carter, R., Mendis, C., De Zoysa, A., Herath, P. and Mendis, K. (1991) Clustering of malaria infections within an endemic population: risk of malaria associated with the type of housing construction. *American Journal of Tropical Medicine and Hygiene* **45**, 77–85.

Gilgen, D.D., Mascie-Taylor, C.G. and Rosetta, L.L. (2001) Intestinal helminth infections, anaemia and labour productivity of female tea pluckers in Bangladesh. *Tropical Medicine and International Health* **6**, 449–457.

GoL (2007) Powering Progress. Government of Lao (http://www.poweringprogress.org).

Gunawardena, D., Wickremasinghe, A., Muthuwatta, L., Weerasingha, S., Rajakaruna, J., Senenayaka, T., Kotta, P., Attanayake, N., Carter, R. and Mendis, K. (1998) Malaria risk factors in an endemic region of Sri Lanka and the impact and cost implications of risk-factor based interventions. *American Journal of Tropical Medicine and Hygiene* **58**, 533–542.

Hadis, M., Lulu, M., Makonnen, Y. and Asfaw, T. (1997) Host choice by indoor-resting *Anopheles arabiensis* in Ethiopia. *Transactions of the Royal Society of Tropical Medicine and Hygiene* **91**, 376–378.

Jobin, W.R. (1999) Dams and Disease. Ecological Design and Health Impacts of Large Dams, Canals and Irrigation Systems. London, New York: E & F N Spon.

Jobin, W. (2003) Health and equity impacts of a large oil project in Africa. *Bulletin of the World Health Organization* **81**, 420–426.

Keiser, J., Singer, B.H. and Utzinger, J. (2005a) Reducing the burden of malaria in different eco-epidemiological settings with environmental management: a systematic review. *Lancet Infectious Diseases* **5**, 695–708.

Keiser, J., Castro, M.C., Maltese, M.F., Bos, R., Tanner, M., Singer, B.H. and Utzinger, J. (2005b) Effect of irrigation and large dams on the burden of malaria on a global and regional scale. *American Journal of Tropical Medicine and Hygiene* **72**, 392–406.

Keiser, J. and Utzinger, J. (2005) Emerging foodborne trematodiasis. *Emerging Infectious Diseases* **11**, 1507–1514.

Keiser, J. and Utzinger, J. (2007) Advances in the discovery and development of novel trematocidal drugs. *Expert Opinion on Drug Discovery* **2** (Suppl. 1), S9–23.

Kemm, J. and Parry, J. (2004) The development of HIA. In *Health Impact Assessment: Concepts, Theory, Techniques, and Applications* (eds J. Kemm, J. Parry and S. Palmer), pp. 15–23, Oxford University Press Inc., New York.

Kirby, M.J., Green, C., Milligan, P.M., Sismanidis, C., Jasseh, M., Conway, D.J. and Lindsay, S.W. (2008) Risk factors for house-entry by malaria vectors in a rural town and satellite villages in The Gambia. *Malaria Journal* **7**, 2.

Kobayashi, J., Phompida, S., Toma, T., Looareensuwan, S., Toma, H. and Miyagi, I. (2004) The effectiveness of impregnated bed nets in malaria control in Laos. *Acta Tropica* **89**, 299–308.

Kobayashi, J., Somboon, P., Keomanila, H., Inthavongsa, S., Nambanya, S., Inthakone, S., Sato, Y. and Miyagi, I. (2000) Malaria prevalence and a brief entomological survey in a village surrounded by rice fields in Khammouane province, Lao PDR. *Tropical Medicine and International Health* **5**, 17–21.

Krieger, G.R., Magnus, M. and Hassig, S.E. (2004) HIV/AIDS prevention programs: methodologies and insights from the dynamic modelling literature. *Clinics in Occupational and Environmental Medicine* **4**, 45–69.

Lautze, J., McCartney, M., Kirshen, P., Olana, D., Jayasinghe, G. and Spielman, A. (2007) Effect of a large dam on malaria risk: the Koka reservoir in Ethiopia. *Tropical Medicine and International Health* **12**, 982–989.

Lerer, L.B. and Scudder, T. (1999) Health impacts of large dams. *Environmental Impact Assessment Review* **19**, 113–123.

Li, Y.S., Raso, G., Zhao, Z.Y., He, Y.K., Ellis, M.K. and McManus, D.P. (2007) Large water management projects and schistosomiasis control, Dongting Lake region, China. *Emerging Infectious Diseases.* **13**, 973–979.

Lindsay, S.W., Jawara, M., Paine, K., Pinder, M., Walraven, G.E.L. and Emerson, P.M. (2003) Changes in house design reduce exposure to malaria mosquitoes. *Tropical Medicine and International Health* **8**, 512–517.

Listorti, J.A. and Doumani, F.M. (2001) *Environmental Health: Bridging the Gaps*. World Bank Discussion Paper 422, World Bank, Washington, D.C.

Lumsden, D. (1993) Dams, Displacement and Mindful Distress. International Mental and Behavioural Health Project, Center for the Study of Culture and Medicine, Harvard Medical School, Boston.

McCully, P. (2001) Silenced Rivers: The Ecology and Politics of Large Dams. London: Zed Books.

Mercier, J.-R. (2003) Health impact assessment in international development assistance: the World Bank experience. *Bulletin of the World Health Organization* **81**, 461–462.

Montgomery, E., Bennett, J.W. and Scudder, T (1973) The impact of human activities on the physical and social environments: new directions in anthropological ecology. *Annual Reviews of Anthropology* **2**, 27–61.

NTPC (2004) Social Development Plan. Chapter V – Health Impact Assessment and Public Health Action Plan. Nam Theun 2 Power Company; Vientiane, Lao PDR http://www.namtheun2.com accessed April 2008.

Odermatt, P., Nanthaphone, S., Barennes, H., Chanthavysouk, K., Tran, D.S., Kosanouvong, B., Keola, S., Mathouchanh, P., Choumlivong, K., Keoluangkhot, V., Phoumindr, N., Nanthanavone, S., Phrommala, S., Degrémont, A. and Strobel M. (2007) Improving tuberculosis case detection rate with a lay informant questionnaire: an experience from the Lao People's Democratic Republic. *Bulletin of the World Health Organization* **85**, 727–731.

Oesterholt, M.J., Bousema, J.T., Mwerinde, O.K., Harris, C., Lushino, P., Masokoto, A., Mwerinde, H., Mosha, F.W. and Drakeley, C.J. (2006) Spatial and temporal variation in malaria transmission in a low endemicity area in northern Tanzania. *Malaria Journal* **5**, 98.

Rowe, G. and Wright, G. (1999) The Delphi technique as a forecasting tool: issues and analysis. *International Journal of Forecasting* **15**, 353–375.
Scudder, T. (1973) The human ecology of big projects: river basin development and resettlement. *Annual Review of Anthropology* **2**, 45–55.
Scudder, T. (2005) The Future of Large Dams. London: Earthscan.
Sota, T. and Mogi, M. (1989) Effectiveness of zooprophylaxis in malaria control: a theoretical inquiry, with a model for mosquito populations with two bloodmeal hosts. *Medical and Veterinary Entomology* **3**, 337–345.
Soukaloun, D., Kounnavong, S., Pengdy, B., Boupha, B., Durondej, S., Olness, K., Newton, P.N. and White, N.J. (2003) Dietary and socio-economic factors associated with beriberi in breastfed Lao infants. *Annals of Tropical Paediatrics* **23**, 181–186.
Staedke, S.G., Nottingham, E.W., Cox, J., Kamya, M.R., Rosenthal, P.J. and Dorsey, G. (2003) Proximity to mosquito breeding sites as a risk factor for clinical malaria episodes in an urban cohort of Ugandan children. *American Journal of Tropical Medicine and Hygiene* 69, 244–246.
Steinmann, P., Keiser, J., Bos, R., Tanner, M. and Utzinger, J. (2006) Schistosomiasis and water resources development: systematic review, meta-analysis, and estimates of people at risk. *Lancet Infectious Diseases* **6**, 411–425.
Subramanian, S., Manoharan, A., Sahu, S., Jambulingam, P., Govardhini, P., Mohapatra, S. and Das, P. (1991) Living conditions and occurrence of malaria in a rural community. *Indian Journal of Malariology* **28**, 29–37.
Thompson, R., Begtrup, K., Cuamba, N., Dgege, M., Mendis, C., Gamage-Mendis, A., Enosse, S., Barreto, J., Sinden, R. and Hogh, B. (1997) The Matola malaria project: a temporal and spatial study of malaria transmission and disease in a suburban area of Maputo, Mozambique. *American Journal of Tropical Medicine and Hygiene* **57**, 550–559.
Trape, J.-F., Lefebvre-Zante, E., Legros, F., Ndiaye, G., Bouganali, H., Druilhe, P. and Salem, G. (1992) Vector density gradients and the epidemiology of urban malaria in Dakar, Senegal. *American Journal of Tropical Medicine and Hygiene* **47**, 181–189.
Trung, H.D., Van Bortel, W., Sochantha, T., Keokenchanh, K., Quang, N.T., Cong, L.D. and Coosemans, M. (2004) Malaria transmission and major malaria vectors in different geographical areas of Southeast Asia. *Tropical Medicine and International Health* **9**, 230–237.
Urbani, C., Sinoun, M., Socheat, D., Pholsena, K., Strandgaard, H., Odermatt, P. and Hatz, C. (2002) Epidemiology and control of mekongi schistosomiasis. *Acta Tropica* **82**, 157–168.
Utzinger, J., Wyss, K., Moto, D.D., N'Diekhor, Y., Tanner, M. and Singer, B.H. (2005) Assessing health impacts of the Chad-Cameroon petroleum development and pipeline project: challenges and a way forward. *Environmental Impact Assessment Review* **25**, 63–93.
Utzinger, J., Wyss, K., Moto, D.D., Tanner, M. and Singer, B.H. (2004) Community health outreach program of the Chad-Cameroon petroleum development and pipeline project. *Clinics in Occupational and Environmental Medicine* **4**, 9–26.
WHO (2000) Human Health and Dams. World Health Organization, Geneva.
WHO (2002) Prevention and control of schistosomiasis and soil-transmitted helminthiasis: report of a WHO expert committee. *WHO Technical Report Series* No. 912.
World Bank (2007) Lao PDR at a Glance. Development Data. http://devdata.worldbank.org/AAG/lao_aag.pdf accessed April 2008.

10

Raw wastewater use in agriculture; risk versus benefits

Jeroen HJ Ensink and Wim van der Hoek

10.1 WASTEWATER USE IN AGRICULTURE

The use of urban wastewater or sewage in agriculture is seen by many as a vital component of integrated water management to overcome regional and global water scarcity (Scott *et al.* 2004). The use of treated wastewater is currently encouraged by the European Union, though its use in Europe is still limited, especially when compared to water-scarce countries such as Israel where treated wastewater is expected to meet 60% of the country's agriculture water demand (Haruvy *et al.* 1999). In water-scarce regions, the use of wastewater in agriculture will free up, and prevent the contamination of good quality water resources for the use in urban centres and industry. Although the use of treated wastewater in agriculture is encouraged and promoted, the unregulated use of untreated or partially treated wastewater in agriculture is estimated to be much more common (Scott *et al.* 2004). High construction, land and maintenance costs associated with most wastewater treatment technology, rapid and uncontrolled urban sprawl and farmers perceptions are all reasons why an estimated 3.5 to 20 million hectares of agricultural land are currently irrigated with partially or untreated wastewater (Ensink and van der Hoek 2007). Almost all of this is situated in developing countries.

Wastewater use, in the absence of wastewater treatment, tends to result in municipalities reacting in one of two ways:

- to ban wastewater irrigation and thus force farmers from their land; or
- to pretend that the use of untreated wastewater is not taking place.

Both are undesirable, as one deprives often poor farmers of their sole livelihood, while the other ignores a potential risk to public health. There is, therefore, a need to take a holistic, but most importantly, a realistic look at the use of untreated wastewater in agriculture, so that targeted health interventions can be provided until wastewater treatment can be implemented. This chapter, by using Pakistan as a case study, aims to identify the main health risks and benefits associated with the use of untreated sewage effluent in agriculture.

10.2 PAKISTAN

At the time of Pakistan's last census in 1998, its population stood at 132 million, a number which is expected to double by 2030. Pakistan is still a predominantly agricultural society and the agriculture relies heavily on the Indus Basin Irrigation System, which irrigates over 16 million hectares of agricultural land. As a result of the country's semi-arid to arid climate, agriculture without irrigation is virtually impossible. A nation-wide survey into the prevalence and characteristics of wastewater irrigation, revealed that in more than 80% of all cities and towns in Pakistan the use of wastewater in agriculture was practiced. The only prerequisite for the use of wastewater in agriculture was the presence of a centralized sewerage collection system. In all instances, wastewater was used in the peri-urban areas of the towns and cities around the disposal stations. Wastewater treatment plants were rare and found in only 2% of all cities (Ensink *et al.* 2004). In those cities where wastewater treatment plants were found, only a small proportion of generated wastewater was receiving treatment and, in these areas, farmers preferred the use of untreated wastewater, because of its higher nutrient status and lower salinity compared to treated wastewater. The most commonly grown crops with wastewater were vegetables (Plate 10.1) and fodder grasses.

Plate 10.1. Farmers growing spinach with untreated wastewater in Faisalabad, Pakistan (Photo: J. Ensink)

10.3 HEALTH IMPACT EVALUATION

The use of wastewater in agriculture poses both health risks and health opportunities to farmers, consumers and people living in close proximity to wastewater irrigated sites. The International Water Management Institute started a programme in 2000, which aimed to quantify both the risks and benefits of wastewater use in Pakistan. For this purpose the city of Faisalabad was selected for a five year study programme. This city was selected for the following reasons:

- over 2500 hectares over land were irrigated with domestic wastewater; and
- although a waste stabilization pond (WSP) was present, farmers preferred to use untreated wastewater (Plate 10.2).

At the start of the study, different costs (health risks) and benefits were identified for which separate studies were designed.

Plate 10.2. Pipes taking untreated wastewater from a wastewater drain before it enters the WSP (Photo: J. Ensink)

10.3.1 Risks

The World Health Organization (WHO) guidelines on wastewater use in agriculture identified a wide range of potential health risks associated with the use of wastewater in agriculture (WHO 2006). These health risks mainly stem from the high concentrations of human and animal pathogens in wastewater but also from organic and inorganic substances.

The main health risks identified were:

- gastrointestinal infections, especially helminth infections, in farm workers as a result of direct exposure to wastewater;
- gastrointestinal infections in crop handlers and consumers as a result of the handling and consumption of wastewater irrigated produce;
- contamination of drinking-water groundwater sources with nitrates; and
- heavy metal build up in soil and concomitant uptake and accumulation in edible plants.

10.3.2 Benefits

World-wide, only 20% of the total area under agricultural production is irrigated. However, this area is responsible for over 40% of the global food production and irrigated agriculture is therefore seen as essential in feeding an ever expanding world population (Seckler *et al.* 1999). The benefits of wastewater compared to regular irrigation are:

- the increased availability and reliability of irrigation water;
- the presence of macro nutrients (nitrogen, phosphorus and potassium) which reduces the need for chemical fertilizer; and
- the availability of a water source in close proximity to urban markets, which reduces transportation costs and allows for the cultivation of perishable high value crops like vegetables.

10.4 COST-BENEFIT ANALYSIS

For the cost-benefit analysis, data was sourced from the study conducted in Faisalabad from 2000 to 2005. The different subcomponents of this research have been presented separately. In this chapter we present a cost-benefit analysis at four different levels: farmer, consumer, municipality and downstream water users.

10.4.1 Farmer

Untreated wastewater irrigation in Faisalabad has a long tradition. Local farmers report that untreated wastewater irrigation dates from at least the construction of the water-borne sewage system in 1972, although other farmers claimed that their families had used wastewater from one of the city's main outfall drains for over 80 years. Initial anecdotal evidence from some of the older farmers showed that farmers were very aware of the nutrient properties of the wastewater. Farmers claimed that the nutrient value of the wastewater used for irrigation had deteriorated since the municipality banned farmers from keeping cattle in the city, which meant that dung was no longer flushed into the central sewerage system.

A survey among 50 wastewater farmers and 50 regular farmers, revealed a strong preference for the use of wastewater. If given the choice, 86% of farmers currently using regular irrigation water would have opted to use wastewater, while only 4% of the farmers currently using untreated wastewater would have opted to use regular irrigation water (Ensink *et al.* 2008). An overwhelming majority (92%) of all farmers identified the nutrient value, the reliability of wastewater, or both as the key incentives for the use of wastewater in agriculture. Farmers were found to be actively pursuing access and the right to use wastewater and were willing to pay a hefty fee. In Faisalabad, wastewater fees were between 6 and 12 times higher than fees for regular irrigation water. The irrigation water fee ranged from 620 rupees (US$ 10) to 3700 rupees (US$ 62) per hectare per year depending on the amount and the quality (nutrient content) of wastewater; with the highest fees paid for untreated wastewater (Ensink *et al.* 2007).

The research conducted in Faisalabad confirmed farmer views; farmers that used wastewater were able to apply more water and apply it in shorter intervals as compared to farmers who irrigated with regular irrigation water (Figure 10.1).

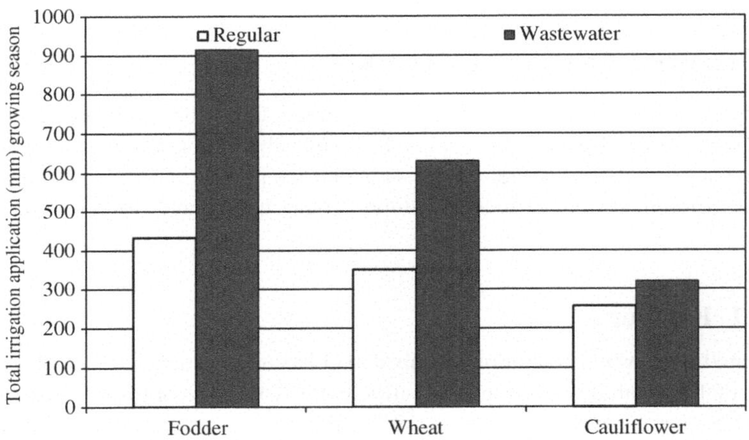

Figure 10.1. Wastewater and regular irrigation water applications (mm) for fodder, wheat and cauliflower in Faisalabad, Pakistan

By using wastewater, farmers were able to meet irrigation quantity requirements for fodder crops (558 mm) and wheat (590 mm), which would have been impossible with regular irrigation water. Regular irrigation water in Pakistan is subject to fixed rotations, where farmers get water once a week, with a fixed timing and quantity, irrespective of the crop grown or its growth stage. Wastewater allowed farmers to be more flexible in their crop choice and allowed farmers to grow vegetables, which have shallow rooting depths and need to be irrigated more frequently.

The nutrient value of untreated wastewater was high with total nitrogen concentrations sometimes exceeding 50 mg/l. Farmers mentioned that some crops could not be grown with wastewater, such as most root crops, while other crops such as wheat and rice sometimes exhibited excessive vegetative growth. However, wastewater was considered ideal for the cultivation of spinach and cauliflower. In Faisalabad, farmers were followed for a year and their fertilizer application was monitored on a weekly basis. The results showed that wastewater farmers predominantly applied fertilizers through the application of wastewater and the chemical fertilizer application was kept to an absolute minimum. When nutrient concentrations and irrigation application were multiplied and chemical fertilizer applications were added, wastewater farmers applied considerably more nutrients than regular farmers (Figure 10.2). Wastewater farmers on average applied 65 kg/ha of chemical fertilizer compared to 530 kg/ha by regular farmers. For the wastewater irrigated site around the WSP this constituted an annual saving of 300 tons of chemical fertilizer or approximately US$ 51,000. If these figures are transposed for the whole of Pakistan, then an annual saving of over US$ 2.5 million could be made.

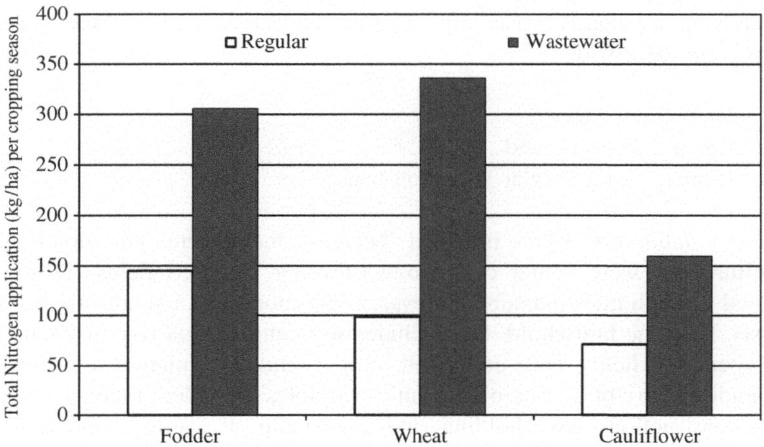

Figure 10.2. Nutrient applications (kg/ha) for fodder, wheat and cauliflower in Faisalabad, Pakistan

Savings in chemical fertilizers made up approximately 30% of total gross margin for wastewater farmers in Faisalabad, with higher cropping intensities and the cultivation of higher valued crops making up the remaining 70%. This meant that wastewater farmers in Faisalabad earned, on average, approximately 35,000 RS (600 US$) per year more than regular farmers.

When asked about the risks of wastewater use in agriculture, 20% of all wastewater farmers mentioned that it could contribute to poor health, although surprisingly none of the farmers could recall a disease episode associated with its usage, often claiming that they had used wastewater for so long that they were immune to it.

The wastewater irrigated site around the WSP was selected for a cross-sectional epidemiological study, investigating the direct health impact of untreated wastewater use in agriculture. The wastewater irrigated site around the WSP comprised approximately 600 hectares and all farmers lived in three communities located in the middle of the wastewater irrigated fields. These three communities together had a population of 4837, of which roughly half lived in households that were involved in wastewater irrigation while the majority of the other half found employment in the textile industry in Faisalabad. A control community located approximately 2.5 kilometres away from the wastewater villages was also selected. This village had a population of 7508 and almost all households were involved in agriculture using regular irrigation water that had previously been found to be free of nematode eggs.

Following a general census and baseline household survey three exposure groups were identified:

- wastewater farmers;
- textile labourers; and
- farmers using regular irrigation water.

Textile labourers were included because they were not involved in agriculture but were living close to wastewater irrigated fields. Data were collected through questionnaire surveys, observations and a single stool sample analysis. Baseline household data included agricultural data (type of activities, time spent in fields and cultivated crops) and information on potential confounding variables. Socio-economic variables included literacy, type of house construction, land holding and ownership of cattle. Sanitation and hygiene related characteristics included the type of water supply, ownership of a motor pump (reflecting good availability of water), presence or absence of a latrine and the use of footwear. The epidemiological cross sectional study conducted in Faisalabad found a low prevalence of helminth infections (Table 10.1) and, in particular, low prevalence of intestinal nematode (hookworm, *Ascaris lumbricoides* and *Trichuris truchiura*) infections (6.5%); the main risks associated with wastewater irrigation. This was surprising, especially given the

the high (>700 eggs/l) concentration of intestinal helminth eggs found in the untreated wastewater. When different confounding variables were controlled for, wastewater farmers showed an increased risk only for hookworm infection (Odds ratio: 23; 95% confidence interval: 5.5 - 96) and *Giardia intestinalis* infection (Odds ratio: 3.3; 95% confidence interval: 2.5 – 4.4). Textile labourers were found in general to have a helminth prevalence in between that of wastewater farmers and regular farmers. Textile labourers were found to have an increased risk of hookworm and *Giardia* infection, though this risk was lower than found for wastewater farmers (Ensink *et al.* 2005; Ensink *et al.* 2006).

Table 10.1. Prevalence of helminth and protozoal infections in wastewater farming, textile labourer and regular farming households

	Wastewater farm households	Textile labourer household	Regular farm households
Helminth			
Ascaris lumbricoides	2.5%	2.2%	0.8%
Hookworm	8.8%	4.4%	0.4%
Trichuris Trichiuria	1.0%	0.7%	0.0%
Hymenolepis nana	13.0%	8.4%	2.5%
Taenia saginata	0.4%	0.1%	0.4%
Other			
Giardia intestinalis	77.4%	72.1%	49.2%
Diarrhoea	5.2%	4.0%	1.9%
Giardia + diarrhoea	4.1%	2.8%	1.3%

Although the epidemiological study found an increased risk of both hookworm and *Giardia* infection in textile labourers and wastewater farmers, it is important to add that both wastewater farmers and textile labourers were of lower socio-economic status and had poorer access to water supply and sanitation than the regular farm households. This was reflected in poorer housing, a lower availability of water and more people practising open defecation. In addition, although all villages had sewage disposal infrastructure, often consisting of a system of open drains, wastewater was often found ponding in the 'wastewater village' (Plate 10.3), while the 'regular' farmer village was found to be free from wastewater year round. Finally, the prevalence of helminth infection found in Faisalabad was low and in line with national figures, while the large majority of the *Giardia* infections was found to be asymptomatic.

Plate 10.3. Wastewater ponding in the streets of one of the 'wastewater' study villages (Photo: J. Ensink)

10.4.1.1 Conclusion

Farmers claimed that wastewater irrigation posed no risk to their health as they claimed to have become immune to the wastewater. The results of the epidemiological survey did not confirm their views, but the direct health impact of wastewater irrigation on farmer health was not dramatic. There was a small difference in the prevalence between regular and wastewater farmers which is unlikely to cause major health problems. Pakistan operates a system of free Basic Health Units where anthelmintic medication[1] is readily available in sufficient quantities. Biannual anthelmintic treatment programmes, combined with medical health camps, would probably eliminate the health problems associated with the direct exposure to untreated wastewater.

It is clear that, for farmers, the benefits of wastewater irrigation far outweigh the risks. Wastewater was found to be more reliable and more flexible allowing

[1] Drugs that act against helminth infections.

farmers to grow more profitable crops, and have a higher cropping intensity, while the nutrient value of wastewater led to large savings in chemical fertilizer cost. The advantage of growing vegetables means that farmers can harvest more than two times per year, and can even play the market and thereby maximize their income. An added advantage of vegetable cultivation is that some produce will be retained for household use, which will contribute to a healthier diet and consequently better health.

10.4.2 Consumers

The nationwide survey in Pakistan showed the importance of wastewater for the cultivation of vegetables. It was estimated that 26% of all vegetables grown in Pakistan were grown using wastewater (Ensink *et al.* 2004). In Faisalabad, weekly market prices of cauliflowers were obtained from the local market committee. The average price of locally produced wastewater-grown cauliflowers was Rupees 703 (SD = 364) against Rupees 1703 (SD 382) for cauliflowers that were transported from other areas of the country. The prices of vegetables were determined solely on the basis of supply and demand forces and it is clear that prices go down dramatically when sufficient local produce enters the market. For the poor segments of the urban population, especially, the lower prices may have positive implications for their vegetable consumption pattern and subsequently their nutritional status. In addition, short transportation and shelf life will have a positive effect on the nutritional quality of fresh produce.

In Faisalabad, the wastewater used for irrigation was of predominantly domestic origin with low levels of organic and inorganic pollutants (Ensink *et al.* 2008). The wastewater irrigated sites in and around Faisalabad exhibited elevated levels of heavy metals as a result of wastewater irrigation, though levels were within permissible limits and no significant uptake by plants was found to have taken place (Ensink *et al.* 2008). The main risk for consumers therefore stemmed from the high *E. coli* (1.8×10^7 cfu/100 ml) and helminth egg (968 eggs/l) concentrations found in untreated wastewater.

To assess the impact of poor quality water use on produce quality, vegetable samples were collected from two farms from April 2004 to March 2005, on a fortnightly basis. At these two farms, vegetables ready for harvest were collected and analysed for *E. coli* and helminth eggs. Vegetables were always sold to the same market salesman; and this market salesman was visited the next morning at his market stall. Here, the vegetables from the same harvest batch as sampled the previous day were purchased and analysed for *E. coli* and helminth eggs. Produce samples were collected from the market between 8 and 10 am to allow sufficient market handling of the produce. Vegetables irrigated by river water were not found to be cultivated in close proximity to Faisalabad.

Although high pathogen concentrations were found in untreated wastewater, these concentrations did not translate into high concentrations on locally grown vegetables. This was probably due to the bed and furrow irrigation technique, which prevents direct contact between wastewater and produce and the local climatic conditions, with high temperatures and low humidity, which promoted rapid die-off of pathogens.

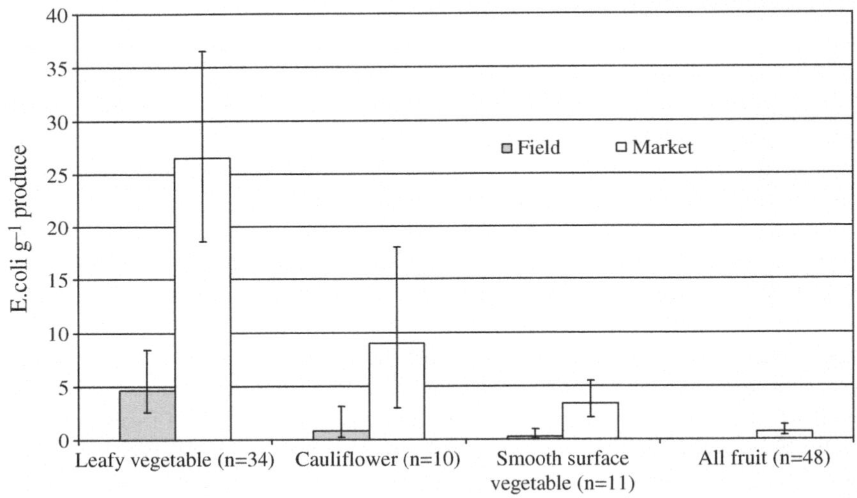

Figure 10.3. *E. coli* concentrations of vegetables and fruit samples collected from wastewater irrigated fields and the local market

During the survey, a total of 110 vegetable samples were collected; 55 at the wastewater irrigated fields and 55 at the market. A total of 8 different vegetables were collected and subjected to microbial analysis. The survey found relatively low concentrations of *E. coli* (geometric mean 1.9 *E. coli*/g) at the farm. Of the different sub-groups of vegetables, the low growing leafy vegetables were found to have the highest concentrations of *E. coli* (Figure 10.3). Vegetables collected from the market were found to have statistically significantly higher concentrations of *E. coli* (geometric mean 14.3 *E. coli*/g) compared to those analysed directly from the field. The difference between field and market was highly significant for each of the vegetable subtypes (Figure 10.3). Helminth eggs on the field vegetables were found at a geometric mean concentration of 0.7

eggs/g, with leafy vegetables showing the highest concentrations (1.1 eggs/g), while cauliflowers had low levels (0.1 eggs/g) and smooth surfaced vegetables were found to be free of helminth eggs. The vegetable samples collected from the market contained, on average, three times more helminth eggs (2.1 eggs/g) compared to the vegetables collected from the field.

The survey also found *E. coli* on fruit samples irrigated with normal irrigation water. The results of the survey suggest that unhygienic post-harvest handling was the major source of produce contamination. This implies that interventions at the market, such as the provision of clean water to wash produce, are better and more cost-effective ways to protect public health than wastewater treatment.

10.4.2.1 Conclusion

Most consumers in Faisalabad and probably all other cities in Pakistan, are unaware of the type of irrigation water used to produce their vegetables. This poses a potential health risk. The unhygienic handling of produce at the local market, however, results in even higher pathogen concentrations on vegetables and fruit than the use of untreated wastewater for irrigation. The overall benefit for consumers therefore consists of a reduced price and increased availability of fresh vegetables.

10.4.3 Municipality

The use of wastewater has posed a dilemma to the local water and sanitation agency (WASA) in Faisalabad. Farmers over the last 10 years have gone to court to gain access to wastewater. The last court case granted farmers the right to use wastewater as farmers had no access to another water source and wastewater was thus their sole livelihood. WASA further acknowledged that there was basically no possible way that the use of wastewater could be stopped as farm inlets, once closed, would be reopened in matter of hours.

The municipality currently operates the WSP at an annual cost of US$ 100,000. The WSP was built with foreign assistance with the objective to provide agriculture with a safe source of water. However, with farmers unwilling to use treated effluent, the WSP runs at a loss. Agreements have been reached with local farmers over wastewater fees and the WASA is renting out land around the WSP to farmers, as a way to generate money. WASA-Faisalabad, like most water supply and sanitation utilities in developing countries, is cash strapped, the high land and water fees that farmers are willing to pay for untreated wastewater offers WASA a way to recuperate money and invest this in improvements in its water supply and sanitation network.

The contamination of groundwater as a result of wastewater irrigation would have been a major concern for the municipal council. The study findings showed high concentrations of *E. coli*, nitrate and nitrites in groundwater. However, in Faisalabad and many other areas of Pakistan, groundwater is unfit for domestic purposes as a result of naturally occurring high salinity.

10.4.4 Downstream water users

As a result of naturally occurring salinity, approximately 50 million people in Pakistan rely on irrigation canals for their domestic water supply, including drinking (van der Hoek *et al.* 2001). In the absence of wastewater treatment, wastewater is disposed of untreated into irrigation canals and rivers, thereby exposing downstream water users to unknown health risks.

10.5 CONCLUSIONS

The Faisalabad case study shows that wastewater use for crop production is a practice with many benefits. It sustains livelihoods of poor peri-urban farming families, contributes to urban food security, helps in solving the urban sanitation problem by preventing pollution of surface water and makes optimal use of the resources (water and nutrients). The health risks associated with wastewater use in agriculture to farmers and consumers of produce can be reduced by proper irrigation water management and implementation of existing public health measures, even when wastewater treatment is not feasible.

10.6 RECOMMENDATIONS FOR SAFE USE

The research in Faisalabad showed a negative impact of untreated wastewater use on the health of wastewater farming families. However, in the absence of wastewater treatment the only alternative to wastewater use in Pakistan is the disposal of untreated wastewater into rivers and irrigation canals. This could potentially pose a much greater health risks as river and irrigation water serve as the only source of domestic drinking water for millions of people in the rural areas of Pakistan, especially in those areas with high natural groundwater salinity. This presents a situation whereby either a small group of wastewater farmers is exposed to a known and relatively easy to contain health risk or a potentially much larger group of downstream water users is exposed to an unknown health risk if wastewater is discharged directly into surface water bodies. To ban wastewater use in Pakistan would be counterproductive and unenforceable, as is acknowledged by local authorities throughout Pakistan. However the current situation where wastewater irrigation is either not recognised or where authorities turn a blind eye, is not sustainable either. Wastewater irrigation makes important contributions to urban food security and farmer livelihoods in Pakistan, however its risks should not be ignored. We propose that the risks that are associated with the use of untreated (domestic) wastewater could be contained through a set of practical and enforceable measures. The first step would to formalize, on a national level, two local regulations which were found to have been agreed between farmers and local authorities throughout Pakistan. These are:

- only produce that will be consumed after cooking can be grown with untreated wastewater; and
- root crops, or other crops grown underground, like potatoes, onions and carrots can not be grown with untreated wastewater.

To minimize the direct health impact of untreated wastewater use on farming families we propose:

- biannual mass treatment of farmers and their children with anthelmintic drugs; and
- general improvements in water supply and sanitation in wastewater villages and local produce markets.

Anthelmintic medication is manufactured in Pakistan and is, thus, readily available and cheap. Rural health workers and Basic Health Units, which provide free medical care, were found to have large stocks of anthelmintic medication available. The availability of a health care infrastructure close to, or in, wastewater irrigated villages in combination with well established wastewater farmer associations offers an excellent opportunity to distribute anthelmintic medication. Children could successfully be targeted though school based (anthelmintic) programmes. Most wastewater farmers belonged to weaker socio-economic groups and the study found lower levels of water supply and sanitation coverage in wastewater villages compared to regular farming villages. Investments in the improvement of local water supply and sanitation might provide greater public health benefits, and could be more cost effective than wastewater treatment. Improvements in local water supply will have the added benefit that more (clean) water will be available at local markets, which will improve post harvest practices and, thus, the quality of the produce. The use of footwear in irrigated fields, often suggested as a suitable preventive measure against hookworm infection, was considered uncomfortable and impractical by farmers and is unlikely to be an acceptable intervention in Pakistan.

10.7 REFERENCES

Ensink, J.H.J., Mahmood, T., van der Hoek, W., Raschid-Sally, L. and Amerasinghe, F.P. (2004) A nation-wide assessment of wastewater in Pakistan: an obscure activity or a vitally important one? *Water Policy* **6**, 197–206.

Ensink, J.H., van der Hoek, W., Mukhtar, M., Tahir, Z. and Amerasinghe, F.P. (2005) High risk of hookworm infection among wastewater farmers in Pakistan. *Transactions of the Royal Society of Tropical Medicine and Hygiene* **99**, 809–818.

Ensink, J.H.J., van der Hoek, W. and Amerasinghe, F.P. (2006) *Giardia duodenalis* infection and wastewater irrigation in Pakistan. *Transactions of the Royal Society of Tropical Medicine and Hygiene* **100**(6), 538–542.

Ensink, J.H., Mahmood, T. and Dalsgaard, A. (2007) Wastewater-irrigated vegetables: market handling versus irrigation water quality. *Tropical Medicine and International Health* **12**(Suppl 2), 2–7.

Ensink, J.H. and van der Hoek, W. (2007) Editorial: new international guidelines for wastewater use in agriculture. *Tropical Medicine and International Health* **12**, 575–577.

Ensink, J.H.J., Simmons, R.W. and van der Hoek, W. (2008) Livelihoods from wastewater: water reuse in Faisalabad, Pakistan. In *International Survey on Water Reuse* (eds B. Jimenez and T. Asano), IWA Publishing, London, UK

Haruvy, N., Offer, R., Hadas, A. and Ravina, I. (1999) Wastewater irrigation – economic concerns regarding beneficiary and hazardous effects of nutrients. *Water Resources Management* **13**, 303–314.

Scott, C.A., Faruqui, N.I. and Raschid-Sally, L. (2004) Wastewater use in irrigated agriculture: management challenges in developing countries. In *Wastewater use in irrigated agriculture: confronting the livelihood and environmental realities* (eds C.A. Scott, N.I. Farqui and L. Raschid-Sally) CABI Publishing, Wallingford, UK.

Seckler, D., Barker, R. and Amerasinghe, U. (1999) Water scarcity in the twenty-first century. *International Journal of Water Resources Development* **15**, 29–42.

van der Hoek, W., Konradsen, F., Ensink, J.H., Mudasser, M. and Jensen, P.K. (2001) Irrigation water as a source of drinking water: is safe use possible? *Tropical Medicine and International Health* **6**, 46–54.

WHO (2006) WHO Guidelines for the Safe Use of Wastewater, Excreta and Greywater. Volume II: Wastewater Use in Agriculture. World Health Organization, Geneva, Switzerland.

11

Concluding remarks

11.1 INTRODUCTION

Significant progress has been made over the past 15 years in establishing a theoretical basis for health impact assessment (HIA), including the development, standardization and refinement of methods, procedures and tools for predicting health impacts of policies, programmes and projects (Parry and Kemm 2004; Scott-Samuel and O'Keefe 2007). Nevertheless, HIA is still at an early stage of development and is rapidly evolving. It therefore varies widely by setting, in both scale and methods employed (Parry *et al.* 2005). While a number of countries in the industrialized world have developed formal mechanisms for assessing health impacts and have effectively institutionalized HIA (Krieger *et al.* 2003; Scott-Samuel and O'Keefe 2007), there is a distinct lack of uniformity in the assessments of new policies, programmes and projects in the developing world. Indeed, a recent systematic examination of the peer-reviewed literature revealed that 94% of the HIA-related publications were conducted in high- and

© 2008 IWA Publishing. *Health Impact Assessment for Sustainable Water Management.* Edited by Lorna Fewtrell and David Kay. ISBN: 9781843391333. Published by IWA Publishing, London, UK.

middle-income countries, whereas only 6% of the published work had an explicit focus on developing country settings (Erlanger *et al.* 2008). To fill this 6/94 imbalance in HIA – worse that the widely known 10/90 gap in health research (i.e. only 10% of the funding in health research is utilized for diseases that cause an estimated 90% of the global burden of disease; Stevens 2004) – represents a formidable challenge. Among others, one objective of this book is actually to narrow the 6/94 gap in HIA by raising awareness among water resources planners and managers regarding the potential of HIA and by illustrating its feasibility through case studies carried out in the developing world. Indeed, Chapters 8–10 focus on HIAs of water resources development and management projects in the developing world, while Chapters 2–7 emphasize water management issues in the United Kingdom and, hence, three of the eight case studies were carried out in developing countries. The pivotal role of HIA when deployed in projects in developing country settings cannot be overemphasized. This relates to perspectives of economics, equity, public health, vulnerable population groups, sustainability and good governance. This holds true especially in the context of natural resources development and exploration, such as water resources development and management exemplified in the aforementioned Chapters 8–10, and also mining, petroleum development and pipeline projects (IPIECA 2005; Singer and Castro 2007; Erlanger *et al.* 2008). In the case of water resources development and management, the evolution of large dam construction around the world, particularly the rapid pace with which large dams have been built in the developing world in recent years (see Chapter 1, Figure 1.1) underscores the pressing need for an HIA framework with effective procedures in developing countries. Indeed, large dam construction for hydropower production and irrigated agriculture and other large infrastructure developments are expected to continue and further accelerate in the developing world. The uneven distribution of health risk, most notably schistosomiasis, malaria and other vector-borne diseases, due to such large dam constructions and irrigation schemes have been documented (Amerasinghe 2003; Steinmann *et al.* 2006; Lautze *et al.* 2007).

It is clear that the preceding chapters in this book reflect HIAs in a diversity in contexts, complexity, geographical boundaries as well as scope and magnitude. The common theme is sustainable water management and examples have been selected from both industrialized and developing countries with studies ranging from the district; i.e. 'housing-development' scale in the United Kingdom, up to a major regional hydro-electric scheme in Southeast Asia. Approaches have ranged from rapid desk studies, mainly based on readily available data, to comprehensive and participatory in-depth investigations which are multidisciplinary in nature, including stakeholder communication and consultation, and involving substantial empirical data acquisition. The latter approach holds promise to strengthen and expand the evidence-base and is characterised by a combination of qualitative and quantitative information. This kind of approach might be warranted for large-scale development projects, particularly when they are implemented in the developing world where baseline health and socioeconomic data are often lacking (IPIECA 2005; Utzinger *et al.* 2005; Singer and Castro 2007).

The heterogeneity in aims, scope and application of HIAs seen in the preceding chapters of this book and elsewhere in the published literature is inherent to the very nature of an HIA, allowing an adaptive approach to health risk assessment and management in specific settings at different levels. This issue alone does not explain, however, the differences that can be observed in the case studies presented here and elsewhere (Kemm *et al.* 2004). These differences suggest a discipline that is indeed at an early stage of its development (Parry and Kemm 2004; IPIECA 2005). Here, parallels are striking with the early emergence of river basin planning (White 1977) when practitioners were seeking to comprehend diverse problems and societal needs across a range of scales and technologies to provide a practical management paradigm and discipline able to accommodate this diversity. The societal need was to provide integrated management systems addressing challenges as diverse as hydroelectric power provision, flood control, eutrophication prevention, management of emerging vector-borne diseases (particularly in populations living in close proximity to tropical impoundments), resettlement of large populations and benefits allocation according to the needs structures within diverse societies (Saha and Barrow 1981). The principles of river basin planning have progressed to continental-scale legal structures in developed nations exemplified by the United States Clean Water Act and the European Union Water Framework Directive which both establish scientific principles of basin-wide planning as well as mechanisms for societal engagement and participation in decision making (Anon 2000).

After 30 years of progress in river basin development, evidence is still lacking, however, of a single unifying paradigm, i.e. reflected in professional practice, suggesting that a global disciplinary approach spanning all development levels is a feasible goal, which has not yet been achieved. In effect, the challenge has been to apply broad principles very differently dependent on the needs structures of specific development projects. An example of the global principle with heterogeneous application might be the concept of 'Water Safety Planning' proposed by the World Health Organization (WHO), which establishes a drainage basin paradigm for potable water quality management (WHO 2004). Here, the principle is universal but the implementation of management interventions and the risks of failure are hugely different in countries with diverse development levels.

HIA is a cognate discipline of river basin planning but one still developing through exploration of different paradigms as practitioners complete specific investigations. Together with other forms of impact assessment, it takes a strong position in the policy framework for integrated water resources management as adopted at the World Summit on Sustainable Development (UNEP 2002). Some of these approaches are becoming more formalised following reviews and research studies (e.g. the DYNAMO-HIA (DYNAmic MOdel for Health Impact Assessment study funded by the EU Public Health Programme 2003–2008 to provide a tool for examination of the effects of EU policies on smoking, obesity, alcohol, cardio-vascular disease and cancer; Anon 2008). Since HIA is currently characterised by heterogeneous innovation and empirical data acquisition, we feel that it provides an excellent platform to learn early lessons from studies carried out in different settings of the world.

Disciplinary growth and development requires continuous collation, critique and evaluation of the evidence-base, hopefully producing constructive insight and progress. Some observations below on emerging issues illustrate this process in the context of HIA.

11.2 CHALLENGES AND EMERGING ISSUES

The promotion, further evolution and ultimately the institutionalization of HIA, possibly integrated with other impact assessments, are faced with a number of challenges and require the further development of a number of HIA attributes. The challenges are linked to the need for a generic framework of procedures that will allow an effective deployment of HIA – singly or in combination with other impact assessments – in different contexts and at different levels. They are all linked to the need for capacity building at the country level. They are listed and briefly covered below.

11.2.1 Creating an enabling environment

From recent HIA capacity building experiences in the countries of the Mekong basin in Southeast Asia it has become clear that the creation of a enabling policy environment is a pre-condition for HIA to take off in an orderly way. Institutional arrangements and the results of human and institutional resources development for HIA will be short-lived in the absence of such a policy.

As a matter of good governance, a national HIA policy needs to be all-encompassing. This implies that the policy formulation process needs to start with considerable advocacy efforts, directed at technical staff and policy makers not only within the health sector but also in a host of other sectors (e.g. agriculture, finance, forestry, urban planning, water resources development, etc.). It is also essential that the plan for a national HIA policy formulation receives political backing from all relevant sectoral ministries and authorities. The process itself follows a pattern of bilateral and multilateral consultations between these ministries and authorities. Part of the review and negotiations will focus on existing policies and legislations for public health, environmental impact assessment (EIA) and natural resources management (including water resources development and management) in order to ensure that the new HIA policy is in harmony with these. This process therefore provides an opportunity to resolve the issue of different levels of integration between HIA and EIA as well as other impact assessments. A possible way forward is to add specific criteria on the issue of integration of different impact assessments in the screening process. It will then be important that consensus is reached on the text of the final draft policy by the council of ministers and that the policy is subsequently decreed by the Prime Minister's Office, so that all government sectors take note and become engaged. In effect, this is an important step in institutionalizing HIA.

Once a national HIA policy has become effective, the procedures can be set in motion for the screening and scoping of proposed development projects, programmes and policies, allowing HIAs to follow their agreed course. The Minister of Health (MoH) can use the policy as a basis to create an HIA unit, establish posts to perform the essential functions and commission the formulation of detailed national HIA guidelines.

11.2.2 Establishing institutional arrangements

The essential functions needed to complete the HIA process successfully were listed in Chapter 1. The regulatory functions among these come under the mandate of the national public health authorities. In order to perform all functions effectively it will be necessary to develop institutional arrangements that support the coordination and collaboration between different sectors and assign specific roles and responsibilities to each.

There is no standard package of institutional arrangements that can be recommended, because the sectoral structures and levels of decentralization vary too much between countries. What is clear, however, is that a strategic alliance between the environment and health sectors, on issues pertaining to EIA and HIA and to disease prevention and environmental management is an important cornerstone. Since equity is another fundamental issue that any HIA, be it of a new project, programme or policy, should address (Kemm and Parry 2004; Chapter 1), it is conceivable that HIA should be integrated also with social impact assessment (SIA) as part of social justice (Scott-Samuel and O'Keefe 2007).

Institutional arrangements can have a number of formats, but the most common is that of a bilateral or multilateral memorandum of understanding (MoU) between ministries and authorities. In addition to roles and responsibilities, such MoUs will also anchor agreements on the allocation of funds that become available for public health management plans, the monitoring procedures (see below) and sanctions in case of non-compliance.

11.2.3 Developing intersectoral decision-making skills

Professional staff working in ministries and authorities usually have not been trained in intersectoral negotiation and decision-making functions, nor have post descriptions and performance appraisal criteria embraced this intersectoral dimension. This crucially important impediment was first comprehensively described by Hunter *et al.* (1993), and is the subject of a task-oriented, problem-based learning course developed by WHO and partners (Bos *et al.* 2003).

In-service training of professionals from different public sectors should target mid-level managers and should build on the knowledge and experience these professionals already bring with them to a training event. This, therefore, calls for an adult learning approach where these professionals address HIA issues in a realistic context that relates to their every-day working experience (Birley *et al.* 1996). The knowledge and skills required to perform optimally in an intersectoral (i.e. multidisciplinary) context should, however, only be developed once the policy framework and the institutional arrangements are in place. At the same time, professional training curricula will have to be reviewed and updated, so that the training of young professionals includes the development of relevant skills.

11.2.4 Ensuring independent quality control

As noted in Chapter 1, one of the challenges in HIA is to prevent biases from creeping into procedures and methods. Lack of objectivity can result from a number of factors. HIA appraisal is a regulatory function controlled by the Government to ensure constant high quality of HIA reports and the ensuing implementation of public health management plans. A brief elaboration on issues that provide the basis for the appraisal criteria for an HIA report follows:

- **Conflicts of interest**. There may be conflicts of interest between the HIA analysts and/or consultants and their clients. The issue of client relationship is most sensitive when the consultants are paid by the project proponent. Conflict of interest usually leads to a rosier picture being presented than the reality warrants.
- **Inappropriate timing and/or duration**. There are two facets which relate to timing. The first deals with the timing of the HIA during the project cycle as a whole. By definition, an HIA must be prospective, and hence should ideally be initiated during the general planning surveys or feasibility studies of a project, in order to maximize opportunities in determining the most cost-effective way to reach the economic and the health and equity objectives during the entire project cycle. Also, the timing of an HIA in relation to other planning and design activities crucially determines its value. Inappropriate timing (i.e. when the construction has commenced) might render the entire study useless. A number of recent publications documented that HIAs of large infrastructure developments were only implemented many years after initiating EIA and SIA; actually HIA only took place shortly before financial close (Utzinger *et al.* 2005; Singer and Castro 2007; Chapter 9). A second facet relates to the cyclical nature of disease transmission and other public health events. For example, the seasonality of disease transmission needs to be considered when planning the collection of primary public health data. Timing issues are usually outside the consultants' and analysts' control.
- **Inadequate budgeting**. Inadequate budgeting can have important repercussions for the objectivity of an HIA. Under-budgeting an HIA is likely to lead to errors, incorrect conclusions and poor recommendations. This issue can be partially resolved by employing consultants with broad experience and expertise.
- **Lack of expertise, experience or skills**. The key health issues identified at the screening stage will have guided the development of the consultants' profiles. It may also have misguided the profiling. Using existing networks to identify suitable candidates can lead to making compromises with respect to the profile of the experts finally selected. There may have been other constraints (e.g. lack of time and financial resources) explaining why a consultant or a group of experts with profiles matching the needs for a sound HIA could not be selected.
- **Lack of access to information**. The level of access to information may be a critical weakness in the HIA process, particularly in developing country settings where baseline health data are notoriously lacking.

> This lack of baseline health and economic data affects objectivity and leads to erroneous predictions. There may be many obstacles to obtaining information. Informants from different ministries may have been reluctant to share reports and data with the consultants. There may have been prejudices, for instance associated with gender or race. Communication may have been hampered by language barriers. Local communities may have provided the consultant or group of experts with information reflecting their expectations rather than reality.

In applying the available HIA methods, the consultants may have arrived at the wrong conclusions concerning hazards, vulnerable communities and health determinants and how these issues change under enhanced pressure of community risk factors, environmental risk factors and institutional risk factors, or in the final synthesis of the different categories.

Some determinants related to an individual hazard may re-enforce one another (e.g. migration of non-immune resettlers into an area where irrigation development has boosted malaria vector populations), others may cancel each other out (e.g. the risk of injury due to the introduction of agricultural machinery will be reduced if there is sufficient capacity within the agricultural extension services to launch an effective training programme).

In the process of drawing up final conclusions, consultants must take every precaution to ensure that these are well-justified and grounded in evidence. Even if the procedure up to this point has been flawless, this final step may still allow bias, prejudice or convention to reduce the overall value of the HIA.

Sound statistics, reference to experience in previous projects (with a clear indication of the similarities and the differences) and a list of assumptions made will allow testing of how variations in the assumptions affect the final conclusions. If variations in the assumptions do not significantly alter the conclusions, then they can be considered robust.

11.2.5 Stakeholder participation

Public and stakeholder engagement remains a central challenge for the HIA practitioner community, but its pivotal role has been emphasised by the World Bank and the International Petroleum Industry Environmental Conservation Association (IPIECA 2005). Lessons from EIA suggest a hierarchy of public engagement characterised by: (i) apathetics; through (ii) spectators; to (iii) the gladiators (Kasperson and Breitbart 1974). While public participation in decision-making is now a legal requirement of parallel river basin legislation such as the EU Water Framework Directive and the Environmental Impact

Assessment Directive as amended by the Public Participation Directive (Anon 2005, 2006a), it would be optimistic to suggest that the professionals managing major projects have developed a participation model or process which ensures a role and real decision-making emancipation for even the 'gladiator' element of society let alone the larger group of apathetics and spectators. In industrialized countries, modern communication tools, such as the Internet, provide excellent opportunities for top-down communication of the information which is commissioned, interpreted, filtered and 'posted' by the professionals. Additionally, these communication tools provide scope for public inputs at all stages of the project development process. In developing country settings, particularly in poor rural and deprived urban settings, the Internet as a medium for information exchange is limited by lack of universal access, as has been documented in the case of the Chad-Cameroon petroleum development and pipeline project (Utzinger *et al.* 2005). Moreover, true and fruitful public engagement is dependent on the timing, quality and dissemination of information from the professionals managing the participation process and this communication step can be expensive, politically contentious and is seldom free of bias (Shepherd and Bowler 1997; Jobin 2003; Leonard 2003; Utzinger *et al.* 2004).

Potential hurdles and resultant challenges to effective public participation range from a lack of agreed objectives for the process within the professional community managing the process through desire to facilitate decision-making power for engaged public participants. With regard to lack of agreed objectives, at the minimalist level, the professionals' intention can be to produce a large number of '*well informed apathetics*' who would be unlikely to inconvenience the professionals with unwanted inputs to the management process. As one moves along this spectrum, a series of difficult and complex issues must be addressed such as: how representative are the 'engaged' participants?, or is the system being manipulated by special interest groups? Another issue to be addressed is how 'public' can the information sharing be where contentious and commercially sensitive information is involved? Perhaps the last word on issues related to stakeholder participation should be left to Beck (1985) who suggested that "*the real issue is who is deciding what the issues are to be.*"

11.3 ATTRIBUTES

11.3.1 Quantification

Cole and colleagues (2005) state: "*If, as is often the case, sufficient information is available to support quantitative prediction of impacts for one outcome but not others, the analysis needs to be carefully structured and presented so that non-quantified impacts are still thoroughly considered and communicated.*" This statement encapsulates a central challenge in HIA. It is tempting to provide more weight to the quantitative estimate of health impact possibly derived from, for example, quantitative microbial risk assessment (QMRA) when, in reality, such quantification may simply be presented because it is possible to measure one pathogen but not others in environmental matrices, or dose-response relationships are available for certain pathogens but not others. Furthermore, the confidence intervals on such estimates may be so wide (and often unknown) as to render the apparently precise quantification at best semi-qualitative. Thus, the HIA practitioner is responsible to consider, weigh and effectively communicate a spectrum of information which will generally span qualitative and quantitative elements of varying importance to the professional and lay recipients. This aspect of varying levels of uncertainty would seem to call for the development of a form of sensitivity analysis in HIA, similar to what is common practice in economic evaluations. Hence, those parameters where there are inherent uncertainties, could be subjected to a series of one-way or multi-way sensitivity analyses (Gold *et al.* 1996; Drummond *et al.* 1997).

11.3.2 Data needs and availability

HIAs have a range of data needs. While good baseline health data may be readily available in developed countries, this is generally not the case in developing country settings. Even where baseline health data are available it may not be possible to obtain them in a disaggregated manner to explore baseline conditions of different vulnerable groups.

In the United Kingdom-based research, reported in Chapters 3 to 7, extensive use was made of non-fatal accident data derived from a sample of home and leisure accidents serious enough to warrant a visit to a hospital accident and emergency department (Home and Leisure Accident Surveillance System). Unfortunately the most recent data were gathered in 2002 and there are no plans to update the database.

With respect to baseline health data, experience has shown that these are often confused with the actual health impact. There are ample examples of 'HIAs' that present the current, static situation as the health impact, rather than the dynamic process of predicted changes in health determinants that will eventually result in a changed community health status. Such flawed HIAs are often also the ones that focus on the strengthening of health services as a response to the health problem, and do not consider options for design and operational changes in the project itself early on in the life cycle of a project.

11.3.3 Vulnerable groups

Vulnerable groups require special consideration within HIAs, as they often suffer a disproportionate impact. The vulnerable groups may include the very young, the elderly, pregnant women, people with poor health and pre-existing disease (especially the immunocompromised), certain occupational groups and some religious groups. These groups may be vulnerable due to inherent characteristics (e.g. age and gender), differential exposure (e.g. farmers to pesticides), behaviour (e.g. lack of belief in health services) or lack of resources (e.g. poor access to health care).

Whenever possible, vulnerable groups need to be taken into account, but data are not always available to allow this to be done adequately. In QMRA, for example, most dose-response relationships are based on healthy volunteers (often young adult men), thus the relationship may not be appropriate for children and adolescents, the elderly or women of childbearing age. In other areas (in some countries at least) good data exist which allow the exploration of likely impacts by age (e.g. in relation to drowning, where marked age differences are seen, with young children at highest risk – Fewtrell and Kay 2008).

Behaviour, such as deliberate exposure to floodwaters, lack of protective clothing for farmers using agrochemicals and recycling chemical containers to collect drinking-water, may well alter individual risk markedly, and where behaviours are widespread they will also alter the population risk. Establishing the prevalence of such behaviours is difficult, even where there is an opportunity to conduct additional field work to obtain new baseline health data.

The 'Source-Pathway-Receptor' approach and paradigm is an internationally accepted toxicological concept which has served well and provides an invaluable tool for quantification of both population effects and the impacts on vulnerable groups. It also leads to contextualisation of feasible mitigation scenarios through reduction in source, attenuation along the pathway and/or protection of the key receptor(s). Such concepts are dynamic and have the potential for accommodating increased vulnerability from, for example, project-induced changes in pathogen risk due to climate change, greywater reuse or impoundment.

However, it important to note that the HIAs covered in this book report 'population' not 'perceived individual' level risks, principally because the former is the key metric for the policy community, whereas the latter is the key consideration for impacted individuals and is hence the driver of behavioural change. This issue is one of the most difficult aspects to model and, thus, to predict its impact on HIA outputs and accuracy.

11.3.4 Multi-disciplinary nature

Many of the challenges in HIA of water resources development and management are only addressed through integrated multi-disciplinary working between professional teams. Chapter 9 in this book provides an example of good practice in this area. However, integration of what Saha and Barrow (1981) termed the *technocentric* and *ecocentric* development paradigms is often more difficult to achieve than to describe even within relatively homogeneous societal structures of, for example, a development in one country. At the same time, health professionals should be sensitive to the possibility that an unnecessarily high profile for HIA may be perceived as a health-centric paradigm and may be counterproductive in the essential multidisciplinary context. In a global context, such integration must be set into the very diverse 'need structures' of specific societies.

11.3.5 Monitoring and surveillance

Monitoring and surveillance is a critical attribute once an HIA has been completed and the resulting public health management plan gets underway, which needs to be coordinated by the health authorities. There are three complementary forms of monitoring and surveillance: compliance verification, health status monitoring and evaluation of measures.

Any public health management plan is multisectoral in nature and will assign responsibilities for safeguards and mitigating measures to different sectors according to their comparative advantages. There is, therefore, a need to ensure that all stakeholders comply with the responsibilities assigned to them and deliver measures to specified levels of quality. The framework of institutional arrangements must provide incentives and sanctions to make the process of verification work.

The health status of the vulnerable groups will require monitoring throughout the construction and early operation phases. The data collected should be checked against the health status scenario developed by taking together the hazards/risks identified and the safeguards and measures recommended under the HIA. Unforeseen adverse health impacts will need immediate action by the health authorities and should be considered for remediation through design and/ or operational modifications.

Evaluation of the safeguards and measures is not only important for the fine-tuning of public health management in the project under scrutiny, but also to construct an expanding database of measures with proved value under specific circumstances. This will facilitate the formulation of an effective public health management plan based on HIAs of development projects in similar settings.

11.3.6 International organizations

Many medium to large water resources development projects will require financing from external sources, through bilateral and multilateral mechanisms. The institutions providing grants, loans or guarantees have their own policies and decision criteria, including, increasingly, an HIA component. Examples include the guidelines by the Asian Development Bank (Peralta and Hunt 2003), the guidance on HIA in the World Bank's Environmental Assessment Source Book (1997) and the recently published performance standards of the International Finance Corporation (IFC), the private sector lending arm of the World Bank group (IFC 2007). Among the eight standards published, IFC Performance Standard 4 now addresses community health, safety and security.

IFC is likely to become influential as it may be emulated by other multilateral development banks and by private sector banks increasingly subscribing to the so-called 'Equator Principles' (Anon 2006b). With regard to these principles, we applaud their institutionalization, but feel that there is currently insufficient mention of health, and hence we conjecture that the principles might be further amended in order to more explicitly address health issues. The preamble of slightly revised principles, we shall call them 'Equator Principles Plus', might read as follows (emphasis added):

> "*Project financing, a method of funding in which the lender looks primarily to the revenues generated by a single project both as the source of repayment and as security for the exposure, plays an important role in financing development throughout the world. Project financiers may encounter social, environmental* ***and health*** *issues that are both complex and challenging, particularly with respect to projects in the emerging markets. The Equator Principles Financial Institutions (EPFIs) have consequently adopted these Principles in order to ensure that the projects we finance are developed in a manner that is socially responsible and reflect sound environmental* ***and health*** *management practices. By doing so, negative impacts on project-affected ecosystems and* ***the health and wellbeing of*** *communities should be avoided where possible, and if these impacts are unavoidable, they should be reduced, mitigated and/ or compensated for appropriately. We believe that adoption of and adherence to these Principles offers significant benefits to ourselves, our borrowers and local stakeholders through our borrowers'*

engagement with locally affected communities. We therefore recognise that our role as financiers affords us opportunities to promote responsible environmental stewardship and socially responsible ***and healthy*** *development.*"

It is encouraging to note that the EPFIs "*will consider reviewing these Principles from time-to-time based on implementation experience, and in order to reflect ongoing learning and emerging good practice*". Hence, we look forward to seeing health being mentioned more prominently in the future.

Negotiations over financing of projects will have to take place within the framework of such policies. There may be conflicts between national policies and the policies of international agencies. Yet, there may also be opportunities to strengthen national capacities in HIA in the context of such negotiations and the follow-up action that can lead to a more harmonized overall approach to HIA.

11.4 REFERENCES

Amerasinghe, F.P. (2003) Irrigation and mosquito-borne diseases. *Journal of Parasitology* **89** (Suppl.), S14–S22.

Anon (2000) Council of the European Communities. Directive 2000/60/EC of the European Parliament and of the Council of 23 October 2000 establishing a framework for Community action in the field of water policy. *Official Journal of the European Union* L327, 1–72.

Anon (2005) Consultation on the Executive's proposals for implementing Article 3 of the European Directive 2003/35/EC. Scottish Executive (now Scottish Government). Edinburgh, UK, October 79p.

Anon (2006a) Public Participation in Environmental Impact Assessment Consultation Report. Scottish Executive (now Scottish Government). Edinburgh, UK, October 38p.

Anon (2006b) The "Equator Principles" – A Financial Industry Benchmark for Determining, Assessing and Managing Social & Environmental Risk in Project Financing. http://www.equator-principles.com/ accessed 30 April 2008.

Anon (2008) DYNAMO-HIA funded by the EU Public Health Programme 2003–2008 of the European Commission's Directorate General for Health and Consumer Affairs (DG SANCO). http://www.dynamo-hia.eu/object_class/dyhia_workpackages.html accessed 10 April 2008.

Beck, M. B. (1985) *Water Quality Management.* Springer Verlag, New York. 120pp.

Birley, M.H., Bos, R., Engel, C.E. and Furu, P. (1996). A multi-sectoral task-based course: Health opportunities in water resources development. *Education for Health: Change in Training and Practice* **9**(1), 71–83.

Bos, R., Birley, M.H., Furu, P. and Engel, C. (2003) Health opportunities in development. A course manual on developing intersectoral decision-making skills in support of health impact assessment. World Health Organization, Geneva.

Cole, B.L., Shimkhada, R., Fielding, J.E., Kominski, G. and Morgenstern, H. (2005) Methodologies for realizing the potential of health impact assessment. *American Journal of Preventive Medicine* **28**, 382–389.

Drummond, M.B., O'Brien, B., Stoddart, G.L. and Torrance, G. (1997) *Methods for the Economic Evaluation of Health Care Programmes*. Second edition. Oxford, United Kingdom: Oxford. University Press.

Erlanger, T.E., Krieger, G.R., Singer, B.H. and Utzinger, J (2008) The 6/94 gap in health impact assessment. *Environmental Impact Assessment Review* **28**, 349–358.

Fewtrell, L. and Kay, D. (2008) Health impact assessment of water management options: generic guidance note. UK Water Industry Research, London.

Gold, M.R., Siegel, J.E., Russell, L.B. and Weinstein, M.C. (1996) *Cost Effectiveness in Health and Medicine*. Oxford University Press, New York.

Hunter, J.M., Rey, L., Chu, K.Y., Adekolu-John, E.O. and Mott, K.E. (1993). Parasitic diseases in water resources development – the need for intersectoral negotiation. World Health Organization, Geneva.

IPIECA (2005) A guide to health impact assessments in the oil and gas industry. International Petroleum Industry Environmental Conservation Association http://www.ipieca.org/activities/health/downloads/publications/hia.pdf accessed 10 April 2008.

IFC (2007) International Finance Corporation Performance Standards: http://www.ifc.org/ifcext/enviro.nsf/Content/PerformanceStandards accessed 30 April 2008.

Jobin, W. (2003) Health and equity impacts of a large oil project in Africa. *Bulletin of the World Health Organization* **81**, 420–426.

Kasperson, R. and Breitbart, M. (1974) *Participation Decentralisation and Advocacy Planning*. Association of American Geographers, Washington DC.

Kemm, J. and Parry, J. (2004) What is HIA? Introduction and overview. In *Health Impact Assessment: Concepts, Theory, Techniques, and Applications* (eds J. Kemm, J. Parry and S. Palmer) pp. 1–13, Oxford University Press Inc., New York.

Kemm, J., Parry, J. and Palmer S. (2004) *Health Impact Assessment: Concepts, Theory, Techniques, and Applications*. Oxford University Press Inc., New York, pp. 417.

Krieger, N., Northridge, M., Griskin, S., Quinn, M., Kriebel, D., Davey Smith, G., Bassett, M., Rehkopf, D.H., Miller, C. and the HIA "promise and pitfalls" conference group (2003) Assessing health impact assessment: multidisciplinary and international perspectives. *Journal of Epidemiology and Community Health* **57**, 659–662.

Lautze, J., McCartney, M., Kirshen, P., Olana, D., Jayasinghe, G. and Spielman, A. (2007) Effect of a large dam on malaria risk: the Koka reservoir in Ethiopia. *Tropical Medicine and International Health* **12**, 982–989.

Leonard, L. (2003) Possible illnesses: assessing the health impacts of the Chad Pipeline Project. *Bulletin of the World Health Organization* **81**, 427–433.

Parry, J. and Kemm, J. (2004) Future directions for HIA. In *Health Impact Assessment: Concepts, Theory, Techniques, and Applications* (eds J. Kemm, J. Parry and S. Palmer), pp. 411–417, Oxford University Press Inc., New York.

Parry, J.M., Kemm, J.R., on behalf of all the participants of the Evaluation of Health Impact Assessment Workshop (2005) Criteria for use in the evaluation of health impact assessments. *Public Health* **119**, 1122–1129.

Peralta, G.L. and Hunt, J.M. (2003) *A Primer on Health Impacts of Development Programs*. Asian Development Bank, Manila.

Scott-Samuel, A. and O'Keefe, E. (2007) Health impact assessment, human rights and global public policy: a critical appraisal. *Bulletin of the World Health Organization* **85**, 212–217.

Singer, B.H. and Castro, M.C. (2007) Bridges to sustainable tropical health. *Proceedings of the National Academy of Sciences of the United States of America* **104**, 16038–16043.

Steinmann, P., Keiser, J., Bos, R., Tanner, M. and Utzinger, J. (2006) Schistosomiasis and water resources development: systematic review, meta-analysis, and estimates of people at risk. *Lancet Infectious Diseases* **6**, 411–425.

Saha, S.K. and Barrow, C.J. (1981) *River Basin Planning; Theory and Practice*. Wiley London, UK.

Shepherd, A. and Bowler, C. (1997) Beyond the requirements: improving public participation in EIA. *Journal of Environmental Planning and Management* **40**, 725–738.

Stevens P. (2004) *Diseases of Poverty and the 10/90 Gap*. International Policy Network, London.

Utzinger, J., Wyss, K., Moto, D.D., N'Diekhor, Y., Tanner, M. and Singer, B.H. (2005) Assessing health impacts of the Chad-Cameroon petroleum development and pipeline project: challenges and a way forward. *Environmental Impact Assessment Review* **25**, 63–93.

Utzinger, J., Wyss, K., Moto, D.D., Tanner, M. and Singer, B.H. (2004) Community health outreach program of the Chad-Cameroon petroleum development and pipeline project. *Clinics in Occupational and Environmental Medicine* **4**, 9–26.

UNEP (2002) Outcomes of the World Summit on Sustainable Development, Johannesburg. http://www.uneptie.org/outreach/wssd/postjoburg/wssdoutcomes.htm accessed 10 April 2008

White, G.F. (1977) *Environmental Effects of Large Scale River Developments*. Westview Press, Colorado. USA. 172p.

WHO (2004) *Guidelines for Drinking-water Quality. Volume 1 Recommendations*. World Health Organization, Geneva.

World Bank (1997) Health aspects of Environmental Assessment. *Environmental Assessment Sourcebook, update 18.* http://siteresources.worldbank.org/INTSAFEPOL/1142947-1118039106036/20526309/Update18HealthAspectsOfEAJuly1997.pdf accessed 30 April 2008.

Index

IWA Publishing's authorised EU representative for General Product Safety Regulations is Diane D'Arras, 15 rue Duret, 75116 Paris, France, e-mail: safety@iwap.co.uk.

Printed and bound by CPI Group (UK) Ltd, Croydon, CR0 4YY
16/06/2026
02140361-0001